PROGRESS IN

SKILLS

GEOGRAPHY

DAVID GARDNER

KEY STAGE 3

HODDER
EDUCATION
AN HACHETTE UK COMPANY

Orders: please contact Hachette UK Distribution, Hely Hutchinson Centre, Milton Road, Didcot, Oxfordshire, OX11 7HH. Telephone: +44 (0)1235 827827. Email education@hachette.co.uk Lines are open from 9 a.m. to 5 p.m., Monday to Friday. You can also order through our website: www.hoddereducation.co.uk

ISBN: 978 1 5104 7757 5

© David Gardner 2020

First published in 2020 by
Hodder Education,
An Hachette UK Company
Carmelite House
50 Victoria Embankment
London EC4Y 0DZ

www.hoddereducation.co.uk

Impression number   10 9 8 7 6 5

Year   2024

Cover photo © arquiplay77 – stock.adobe.com

Illustrations by Aptara and DC Graphic Design Limited

Typeset in Museo Sans by DC Graphic Design Limited, Hextable, Kent

Printed and bound by CPI Group (UK) Ltd, Croydon, CR0 4YY

A catalogue record for this title is available from the British Library.

Contents

Contents

15 Climate change

16 Reviewing skills

Learning objectives

▶ To ask geographical questions.
▶ To understand the process of geographical enquiry.
▶ To identify different geographical data.
▶ To interpret a photo of a place, using enquiry questions.

Progress in Geography Skills has been created to help you study like a geographer. A vision of what it is to be a good geographer is shown on the vision statement, flap C, on the inside front cover. This book introduces and progresses your capability to develop a range of geographical skills. It contains different types of geographical data and activities to help you interpret and make sense of this data.

At the heart of geography is the process of geographical enquiry. Throughout the book, there are opportunities to bring your skills together in order to answer enquiry questions (like those shown on photo A). By the time you reach the end of the book, you will be able to use the skills you have progressed in order to carry out a high-level enquiries such as Lesson 15.3 shown in screenshot B. The labels in the screenshot of this lesson show the stages of the enquiry process. In the book, some of the lessons introduce a particular skill, other lessons require you to use a range of skills to conduct geographical enquiries.

A Durdle Door, Dorset coast, UK

DANGER - Rockfalls

Keep Clear of cliffs
Serious risk of injury or death

You remain at risk even at low tide

Enquiry questions

Where is this place?

What is it like?

Why is it like this?

How is it changing?

Who is affected by the changes?

How do you **feel** about this place?

Ask questions

Investigate using different types of geographical data

Communicate the findings

Analyse the data

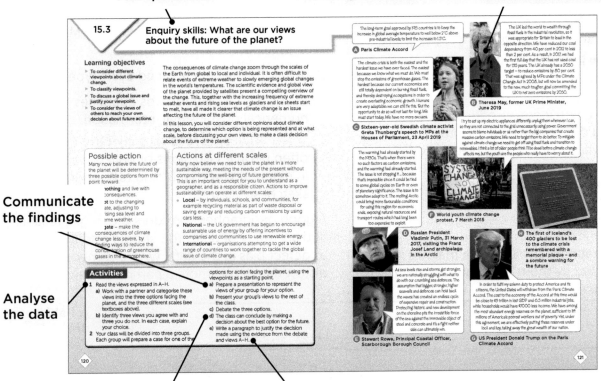

Conclusion – reflect on findings

Make a decision based on evidence

B Stages in the enquiry process for Lesson 15.3, pages 120–121

Activities

1 Draw a table like the one below. Look through the pages of this book and identify three examples of each type of geographical data that you will use to develop your skill.

Add a column to your table to give examples of how you used these types of geographical data in your primary school.

Type of geographical data	Examples from book	Page numbers
Atlas maps		
Ordnance Survey maps		
Photographs – ground level		
Aerial photographs		
Satellite images		
Graphs		
Statistics		
Different points of view		
Newspapers		

2 Geographers use questions to investigate places. Answer the enquiry questions on photo A to describe Durdle Door.

3 a) Look at screenshot B. Write out the elements of the enquiry process as a list of side headings.

b) Think about how you studied geography at your primary school, and add examples beside each heading of how you may have used the elements of enquiry.

4 Look carefully at the vision statement, flap C. Identify and list ways you will be provided with opportunities to progress your geographical enquiry skills in the book.

What is geographical data?

Learning objectives

▶ To recognise the different types of geographical data.
▶ To categorise geographical data.

In Lesson 1.1, you were introduced to the enquiry process and discovered that geographers ask questions when investigating places. To find answers to these questions, geographers collect data (see examples A–G). You can collect data yourself by conducting fieldwork: taking photos, drawing fieldsketches, or carrying out surveys and questionnaires. This is called primary data. Geographers also use data that has not been collected by them. This can be data from textbooks, newspapers, websites or published data, such as climate figures from the Met Office. This is called secondary data.

Data can be further classified by its type:

● Quantitative data records quantities and numerical or statistical information.
● Qualitative data records subjective qualities, often people's opinions or beliefs, presented in different ways, including text, photos or sketches.

A good geographer considers the origin, reliability and accuracy of data. You should also attempt to detect bias in the data, where someone might distort data to promote a particular viewpoint.

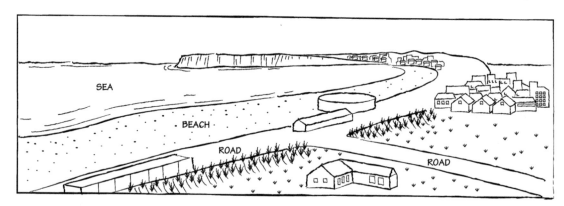

A A fieldsketch drawn by a Year 7 pupil

C Photograph taken by a pupil when conducting fieldwork on the coast

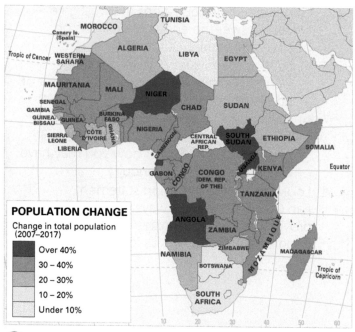

POPULATION CHANGE
Change in total population (2007–2017)

- Over 40%
- 30 – 40%
- 20 – 30%
- 10 – 20%
- Under 10%

B Population change in Africa, 2010, © *Philip's School Atlas*

D A political cartoon about Russian exploration of the Arctic Ocean, 2012

E Spreadsheet showing weather data collected from a school weather station

Date	Temperature max °C	Temperature min °C	Wind av km/hr	Wind max km/hr	Wind direction	Rainfall mm	Sunshine hours	Pressure mb
01/09/2017	18.9	10.6	20.8	25.6	SW	0.3	0.5	1003
02/09/2017	15.5	2.6	0.0	22.7	SW	0.3	1.8	1019
03/09/2017	15.8	2.0	7.6	7.6	SE	0.0	2.3	1010
04/09/2017	12.6	8.0	7.6	18.0	S	2.8	0.5	995
05/09/2017	18.1	9.2	11.4	15.2	S	2.3	1.9	995
06/09/2017	16.1	7.5	2.8	13.3	SE	0.3	0.5	1009
07/09/2017	20.4	8.3	4.7	5.7	SE	0.0	2.4	1026

The stakes are very high. We know these changes are happening, the evidence is incontrovertible. If temperatures continue to rise, it could have a catastrophic effect on the human race. It is not for scientists to dictate to the world, it is for scientists to declare what they see and then allow people to make up their minds.

F Sir David Attenborough's view about climate change

Site	Width (m)	Depth at mid-point (m)	Time for orange to travel 3 m (s)	Speed (m/s)	Other notes
1	10	0.005	88	0.034	Shallow, orange frequently got stuck
2	0.75	0.045	29		
3	1.2	0.03	55		Shallow – orange got stuck
4	1.2	0.085	25.4		
5	3.3	0.04	33		Furthest site from source

G Measurements of a river, collected during fieldwork

Activities

1 What is geographical data?

2 Write definitions for the following types of data: primary, secondary, quantitative, qualitative.

3 Look carefully at the data shown in A–G. For each, first identify whether it is primary or secondary data and then whether it is qualitative or quantitative data.
In each case, write a sentence to explain your choice.

4 Explain why you think a good geographer will check the origin, reliability and accuracy of the data they select to use as evidence to answer an enquiry question.

5 Select one of A–G that you think might be biased, and write a sentence to explain your choice.

6 Look at the vision statement, flap C. Identify and write down the aspects of being a geographer that you think link to using geographical data.

Future learning

You will be provided with opportunities to use a wide range of geographical data to make sense of places throughout *Progress in Geography Skills*. You will also learn how to best present the data that you collect and analyse.

Learning objectives

▶ To understand the difference between lines of latitude and longitude.

▶ To be able to locate places on a physical atlas map of the world, using lines of latitude and longitude.

A geographer develops good locational knowledge of where places are in the world. This is the first enquiry question introduced in Lesson 1.1 – Where is this place? Maps are an important tool used by geographers to do this. Maps provide us with a graphic representation, a bird's-eye view of a place or area, and they can also show us patterns and changes.

Progress in Geography Skills will provide you with opportunities to read and interpret a wide range of different types of maps, including atlas maps. An atlas is a book of maps which show different physical and human features of the world, at a variety of scales, including global, continental and regional or by country. Map B is an atlas map showing the physical geography of the world.

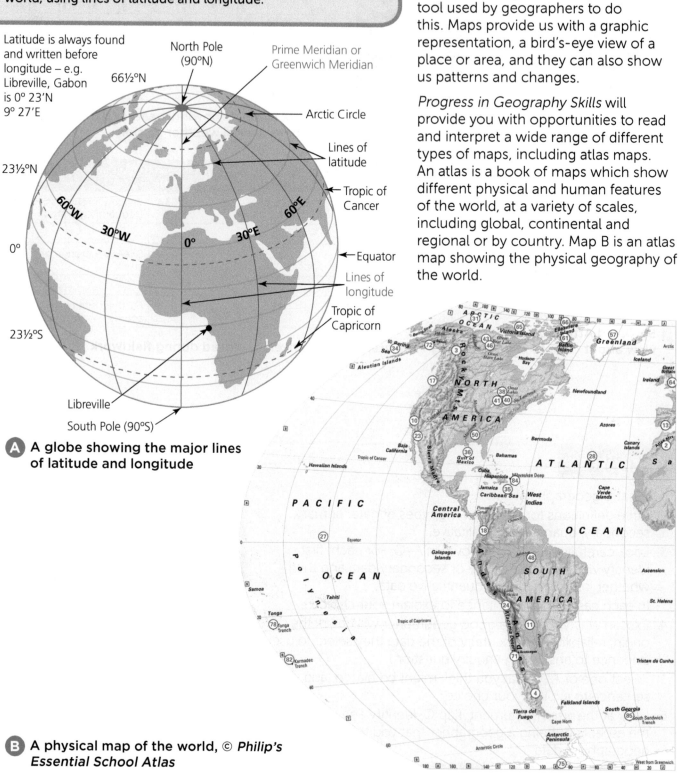

Latitude is always found and written before longitude – e.g. Libreville, Gabon is 0° 23'N 9° 27'E

North Pole (90°N)

Prime Meridian or Greenwich Meridian

66½°N

Arctic Circle

Lines of latitude

23½°N

Tropic of Cancer

60°W

60°E

30°W

30°E

0°

0°

Equator

Lines of longitude

Tropic of Capricorn

23½°S

Libreville

South Pole (90°S)

Ⓐ A globe showing the major lines of latitude and longitude

Ⓑ A physical map of the world, © *Philip's Essential School Atlas*

Lines of latitude and longitude

- Lines of latitude and longitude are imaginary lines drawn on maps and globes, which form a grid for locating places.
- Lines of latitude run east to west around the Earth, numbered in degrees, up to 90°, north and south of the Equator – the 0° line of latitude.
- The Equator divides the Earth into two halves, the Northern and Southern Hemispheres.
- Lines of longitude run north to south from the North to the South Poles. They are numbered in degrees east or west of a line through Greenwich, London, known as the Prime Meridian and numbered 0 degrees. Lines of longitude are numbered up to 180 degrees, to show how far west or east they are from this Prime Meridian.
- Lines of latitude and longitude are shown on A and B.
- Each degree of latitude and longitude is divided into 60 minutes. For example, a place located at latitude seven and a half degrees north can be called 7 degrees and 30 minutes north, written 7° 30'N.

Activities

1 Write definitions of the following terms:
   a) map
   b) atlas
   c) lines of latitude and longitude
   d) Prime Meridian.
2 Diagram A shows five important lines of latitude. Write them as a list and include the number of each line.
3 Look carefully at the physical atlas map of the world, map B.
   a) Write a list of the types of natural features a physical atlas map shows.
   b) Name the continents and oceans of the world.
   c) Make your own copy of the following table, and record the latitude and longitude of the natural features listed.

Natural feature	Latitude and longitude
i) Mount Everest	
ii)	45°N 58°E
iii) Galapagos Islands	
iv)	55°S 67°W
v)	17°N 142°E
vi)	0°S 50°W
vii) Bay of Bengal	
viii)	43°N 34°E
ix) Mount Kilimanjaro	
x)	60°N 59°E

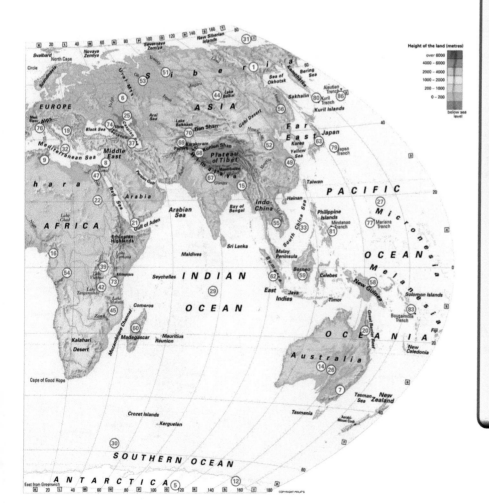

Learning objectives

▶ To use the contents and index of an atlas.

▶ To be able to describe the absolute and relative location of places.

▶ To be able to locate places on a political atlas map of the world, using lines of latitude and longitude.

Geographers use an atlas as a starting point to understand what places are like. At the front of an atlas, the contents page lists where you can find different maps, such as continent or country maps, or thematic maps on climate or environmental regions. The index, at the back of the atlas, will help you find a particular place. Places are listed in alphabetical order. The entry for each place listed in the index has the page number of where it is in the atlas, and its latitude and longitude. See index A.

One way of locating a place is by describing its absolute location, the exact spot it can be found on the globe. This can be done using latitude and longitude when using atlas maps, and grid references when using Ordnance Survey maps. You can also describe the location of a place in relation to other places, by its relative location. This refers to the distance and direction from one place to another. The use of place names, landmarks and regions helps to specify the relative location of one place from another. For example, Cairo is located in the north of Egypt, along the River Nile. Describing the relative location of a place helps to orientate yourself in space and develop an awareness of the world around you.

Ⓐ **Section of an index from** *Philip's Essential Atlas*

C	Place in alphabetical order	Latitude	Longitude	Atlas page number	
Cabinda		5° 0'S	12° 30' E	51	6D
Cabo Verde ■..........		16° 0'N	24° 0'W	70	4H
Cáceres		39°26'N	6°23'W	33	13E
Cader Idris		52°42'N	3°53'W	16	4C
Cádiz		36°30'N	6°20'W	33	14E
Caen		49°10'N	0°22'W	33	8H
Caernarfon		53° 8'N	4°16'W	16	3B
Caerphilly		51°35'N	3°13'W	17	5C
Cágliari		39°13'N	9° 7'E	36	5B
Caha Mountains		51°45'N	9°40'W	19	5B

Political maps

Map B is a political map of the world from an atlas. A political map shows the boundaries of countries, states or counties, as well as the location of major cities. Different colours are used to distinguish between different countries. A border is a real or artificial line that separates countries or states. These boundaries are sometimes visible on the ground, by a sign or even a fence or wall, such as the boundary between the USA and Mexico.

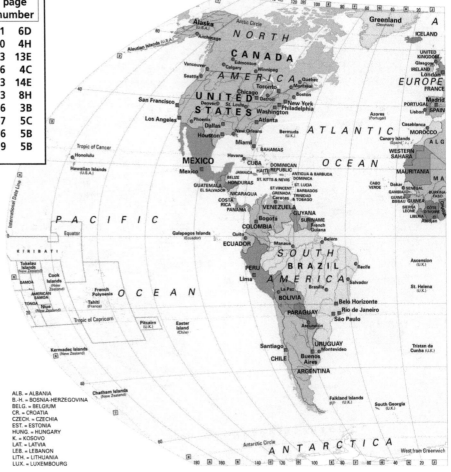

ALB. = ALBANIA
B.-H. = BOSNIA-HERZEGOVINA
BELG. = BELGIUM
CR. = CROATIA
CZECH. = CZECHIA
EST. = ESTONIA
HUNG. = HUNGARY
K. = KOSOVO
LAT. = LATVIA
LEB. = LEBANON
LITH. = LITHUANIA
LUX. = LUXEMBOURG

Activities

1 Using a copy of your school atlas, turn to the contents page. Note down on which page you can find:

   a) the mountains of Asia
   b) the cities of Europe
   c) the climate of Africa
   d) the rivers of South America
   e) a population map of the world.

2 Turn to the index of your school atlas to locate the places shown in index A. Label them on an outline map of the world.

3 Using the index of your school atlas, find the latitude and longitude, atlas page numbers and countries for the following cities – Beijing, Addis Ababa, Dubai, Cairo, Los Angeles. Record your results in a table.

4 Describe what the absolute and relative location of places means.

5 a) Look carefully at the political atlas map B and use it to copy and complete the table below.

   b) Describe the relative location of Singapore and Moscow.

Human feature	Latitude and longitude
i) Cape Town	
ii)	52°N 104°E
iii) Singapore	
iv)	23°S 43°W
v)	40°N 74°W
vi)	33°N 44°E
vii) Sydney	
viii)	64°N 22°W
ix) Moscow	
x)	49°N 123°W

6 Compare the world political atlas map B, with the world physical atlas map B in Lesson 1.3.

   a) Write three sentences to identify the different geographical features these two maps show.

   b) The physical atlas map labels 86 different locations around the world. Compare any 10 of these locations with the political atlas and name the country they are located in. Write your answers as a list, for example 48: Brazil.

M. = MONTENEGRO
MACED. = MACEDONIA
MOLD.= MOLDOVA
NETH.= NETHERLANDS
SERB. = SERBIA
SLO. = SLOVENIA
SLOV. = SLOVAKIA
SWITZ. = SWITZERLAND
U.A.E. = UNITED ARAB EMIRATES
U.K. = UNITED KINGDOM
U.S.A. = UNITED STATES OF AMERICA

B World political map, © *Philip's Essential School Atlas*

Learning objectives

▶ To understand scale.
▶ To understand how scale is shown on a map.
▶ To understand that different scales of map can be used for different purposes.

The Ordnance Survey (OS) is the national mapping agency for Great Britain. It produces up-to-date and accurate maps of the country, at a range of scales for people and businesses that use maps in different ways. If you are to make progress as a geographer, you need to be able to use and read OS maps.

What is a map scale?

The scale of a map shows how much you would have to enlarge your map to get the actual size of the piece of land you are looking at. If, for example, your map has a scale of 1:25 000, this means that every 1 cm on the map represents 25,000 cm or 250 m on the ground. This means that every 4 cm on a map at this scale is equal to 1 km on the ground. A large-scale map shows a lot of detail for a place, but covers a small area. A small-scale map shows a larger area, but less detail. The atlas maps you studied in Lessons 1.3 and 1.4 are examples of small-scale maps.

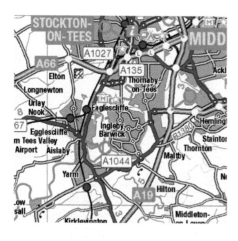

A 1:200 000

0 1 2 3 4
Kilometres

0.5 cm to 1 km

B 1:50 000

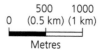

500   1000
0   (0.5 km) (1 km)
Metres

2 cm to 1 km

C 1:25 000

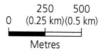

250   500
0   (0.25 km)(0.5 km)
Metres

4 cm to 1 km

OS maps A–E show the town of Yarm at five different scales.

The scale of a map is shown in three different ways on each map:

- As a ratio, for example 1:50 000 – this means that 1 unit on the map is 50,000 times larger on the ground.
- As a line – called a linear scale.
- As a statement – for example, 1 cm = 500,000 cm on the ground, or 0.5 km.

Activities

1 What is the Ordnance Survey?
2 What is scale?
3 Why do we draw maps to scale?
4 Explain the three ways scale is shown on a map and provide examples from OS maps A–E.
5 Draw linear scales for the following other OS scales of map:
    a) 1:25 000
    b) 1:1 000 000.
6 Look carefully at OS maps A–E of Yarm.
    a) Which map and scale covers the largest area?
    b) Which map and scale shows the most detail for Yarm?
    c) Which maps are small-scale and which are large-scale?
    d) Identify and explain which scale of map would be best suited to the following:
        i) an architect designing a new building in Yarm
        ii) driving to Yarm by car
        iii) a family planning a day out in Yarm.
7 Visit the Ordnance Survey website: www.ordnancesurvey.co.uk. Find and record the names for each scale of map shown on maps A–E.

D 1:10 000

0   100   200
Metres

10 cm to 1 km

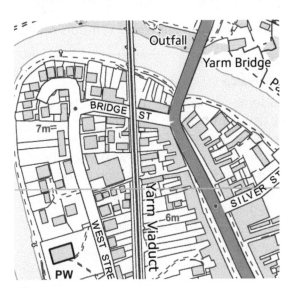

E 1:5000

0   50   100
Metres

20 cm to 1 km

Learning objectives

▶ To understand how and why the OS uses symbols on maps.

▶ To be able to locate places and features on OS maps using four- and six-figure grid references.

OS map symbols

At first glance, an OS map can seem confusing. It is a jumble of lines, shapes, colours and letters. But an OS map contains a massive amount of data about places, and it uses a special language to store this information – symbols. These symbols include patches of colour, lines, drawings or abbreviations (see A), and each one is explained on a special key called a legend (see pages 129–130). Each scale of OS map has a slightly different set of symbols.

The OS symbols are designed to be simple, often looking like the features they represent. This means things can be quickly and easily recognised. As you get used to reading and interpreting different OS maps you will start remembering what these symbols mean, as you learn the language of maps.

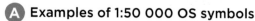

A Examples of 1:50 000 OS symbols

B Examples of features shown as symbols on OS maps

Activities

1 Write a sentence to explain why the Ordnance Survey uses symbols on its maps.

2 Look back at the OS maps of different scales of Yarm in Lesson 1.5. Explain why the 1:5000 map E uses far fewer symbols than maps A–C.

3 Using the legend for 1:50 000 OS maps on pages 129–130:
   a) Identify what the symbols shown in A represent. Be careful, some actually show more than one symbol.
   b) Identify and draw the symbols for the features shown in photos B.
   c) Draw and label five other OS symbols that look like pictures of what they represent.
   d) List five symbols that are letters, and what they represent.
   e) Draw and label three symbols that are lines.
   f) Draw three symbols that are patches of colour.

4 Look at map-flap D of the Swanage area.
   a) Find the small village of Studland. What symbols can you see in and around this village that suggest that tourists visit the area?
   b) Find the settlement of Corfe Castle. Name and identify symbols that suggest this village is of historical importance.
   c) Visit the Ordnance Survey website: www.ordnancesurvey.co.uk. Click on the Education tab at the top of the page. Once at the Education home page, click on Resources for teachers, and then scroll down to Learning map symbols. You can download your own copies of the Landranger 1:50 000 legend and the 1:25 000 Explorer legend. You could save these as a pdf file, and print out a copy of the symbols to keep in your book, to help you read OS maps.

How to locate places on OS maps

OS maps are covered in a series of numbered blue lines that make up a grid. These grid lines allow you to identify the location of a place on the map by giving a unique number known as a grid reference. The vertical lines are called eastings, because they increase in value eastwards on the map. The horizontal lines are called northings, because they increase in value northwards on the map.

The grid reference is always from the bottom left-hand corner of the grid square you are locating. Four-figure grid references only allow you to locate a square kilometre, which is a large area and not very accurate. By imagining each grid square is divided into 100 smaller squares, a six-figure grid reference is more accurate, with each small square representing 100 m.

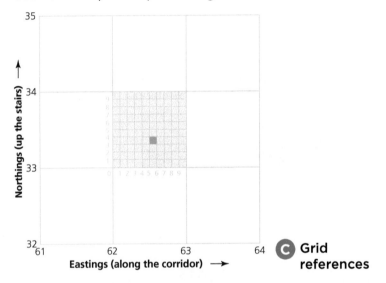

C Grid references

Four-figure grid references

To give a four-figure grid reference for the shaded square:

- Read along the easting line until you reach the bottom left corner of the shaded square. Read off the number: **62**.
- Now read along the lines to the side of the numbered line at the bottom of the shaded square: **33**.
- These numbers combined provide the four-figure grid reference: **6233**.

Six-figure grid references

To locate a feature more precisely within a square, such as the small green shaded square shown, we can use a six-figure grid reference:

- First, imagine that each grid square is divided into tenths (as shown on the grid).
- Read along from square 62 to count the tenths. There are 5. Read off the number: **625**.
- Now read up from square 33 to count the tenths. There are 3. Read off the number: **333**.
- These numbers combined provide the six-figure grid reference: **625333**.

Activities

5 Write a sentence to explain what grid references are.

6 Write two sentences to explain the difference between eastings and northings.

7 Read the instructions on how to use four- and six-figure grid references in C.

   a) Work with a partner. One of you explains, using the grid squares in C, how to give a four-figure grid reference, and then your partner explains how to do a six-figure reference.

   b) Now repeat the process on map-flap D, for Corfe Castle.

8 Look at map-flap D of the Swanage area.

   a) Write down the four-figure grid references for the following:

     i) Studland village

     ii) Durlston Head

     iii) Worth Matravers

     iv) Studland Heath.

   b) Write down the six-figure grid references for the following:

     i) railway station at Swanage

     ii) church with a tower at Worth Matravers

     iii) public house at Corfe Castle

     iv) youth hostel at Swanage

     v) visitor centre at Studland Bay.

   c) Write down what is located at the following six-figure grid references

     i) 017773   iv) 033850

     ii) 015793   v) 997825.

     iii) 988842

How can I measure the distance and direction between two places on an OS map?

Learning objectives

▶ To be able to measure the straight distance between two places on an OS map.

▶ To be able to measure curved distances between two places on an OS map.

▶ To recap the points of the compass and direction.

One of the many ways we can use maps is to find out how far one place is from another. You can do this using a piece of paper to measure the distance on an OS map, and the linear scale.

In this lesson, you will use map-flap E, an OS map of the Cockermouth area. This is a 1:50 000 map. This means that 1 cm on the map is equal to 50,000 cm on the ground (2 cm on the map represents 1 km on the ground). A linear scale is provided on the map-flap.

The instructions given below describe how to measure the distance between the public house at the village of Dovenby, along the A-road and minor road, to the public telephone at the village of Bridekirk.

A How to measure the straight-line distance (note: these photos are not shown to scale)

Step 1: Lay the strip of paper on the map between the two points to be measured.

Step 2: Mark the positions of the two places – A and B.

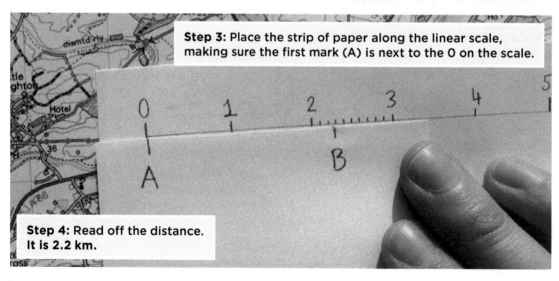

Step 3: Place the strip of paper along the linear scale, making sure the first mark (A) is next to the 0 on the scale.

Step 4: Read off the distance. It is 2.2 km.

The distance along a road between two places with many bends along the route is longer than the straight-line distance. To measure this more realistic and accurate travel distance, you need to move the piece of paper to take each curve into account.

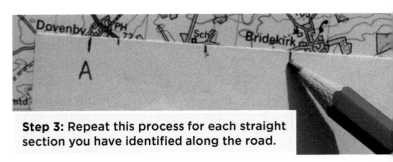

C The compass rose

B How to measure the curved distance
(note: these photos are not shown to scale)

Step 1: Look at the road to be measured and break it down into straight sections. Lay the strip of paper along the first straight section and label it A on the paper.

Step 3: Repeat this process for each straight section you have identified along the road.

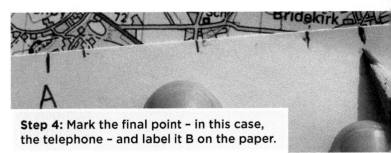

Step 4: Mark the final point – in this case, the telephone – and label it B on the paper.

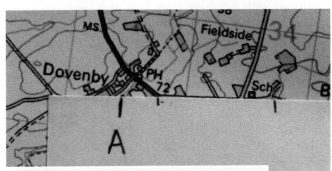

Step 2: Pivot the paper at this first mark, so that the paper now lies along the next straight edge, and add another mark at the end of this next straight section.

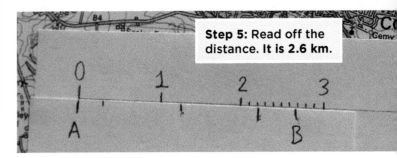

Step 5: Read off the distance. It is 2.6 km.

Activities

1. In your primary school, you will have been introduced to, and used, the points of the compass, shown in C. Make your own copy of the compass rose and add the full name of the direction for each point.

2. Look carefully at map-flap E of Cockermouth.
   a) What are the six-figure grid references for the public house at the village of Dovenby and the public telephone at the village of Bridekirk?
   b) What direction is the public house from the public telephone?

3. Measure the straight-line distance between the following places on map-flap E, following the instructions provided in A:
   a) the church with a tower at Brigham 086309 to the telephone at Eaglesfield 094282
   b) the A66 bridge over the River Cocker 119294 to the river at 150262.

4. Now measure the curved distances:
   a) along the road between the two locations in Question 2a).
   b) along the course of the River Cocker between the places in Question 2b).

Learning objectives

▶ To understand how height is shown on OS maps.

▶ To identify contour patterns.

Ordnance Survey maps are topographic maps – they give a two-dimensional representation of a three-dimensional land surface to show the shape of the Earth's surface. On OS maps, this is done in three ways:

.144

- **Spot heights** – marked by a black dot with the height in metres written next to them.
- **Triangulation pillars** – shown by a blue triangle with a dot in the middle and the height marked next to it. (On the ground, these pillars are positioned on a hill top with a view overlooking the surrounding countryside. They were once used by OS surveyors to measure the land.)
- **Contour lines** – thin brown or orange lines that join together places of the same height. Some contour lines have their height above or below sea level written on them. Contour lines are usually drawn at 10 m intervals on a 1:50 000 scale map and at 5 m intervals on a 1:25 000 scale map. At regular 50 m intervals, the contours are printed with a thicker line. The numbers on contour lines are always displayed in ascending height, so if the numbers increase it is an uphill slope, and if they decrease it is a downhill slope.

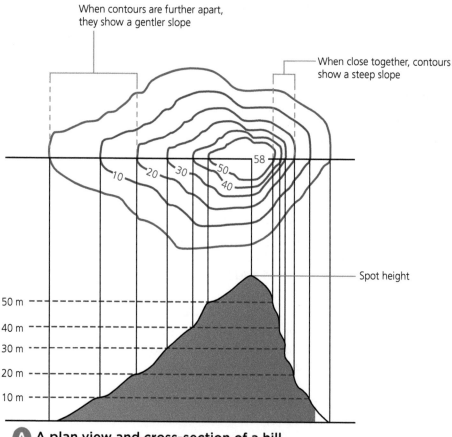

A A plan view and cross-section of a hill

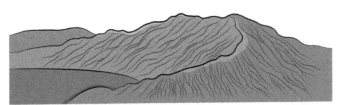
Ridge – steep-sided tongue of upland shown by narrowly spaced contour lines on either side

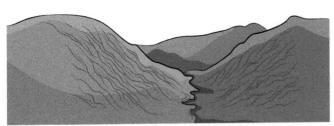
Valley – shown by contour lines pointing upstream, often V-shaped

C Visualisation of two landforms in 3D

B A contour pattern

Visualising contour patterns in 3D

The ability to read and understand the shape of the land from a map is an important skill for a geographer to learn. It takes practice to read a map and recognise contour patterns well enough to be able to visualise what the landscape looks like in three dimensions.

To help you imagine a contour pattern in 3D, diagram A shows the link between the shape of a hill and the contours representing it on the map. When the contours are plentiful and close together, the slope is steep. When they are less dense and spaced well apart, the slope is gentle. When there is an absence of contours, the landscape is flat. As you become more skilled at reading OS maps, you will begin to recognise particular landforms from the contour patterns. Map B and sketches C show some basic contour patterns and landforms.

Activities

1 What is a topographic map?
2 Describe and draw the symbol for the three ways height is shown on OS maps.
3 Look carefully at diagram A. Describe the contour patterns that represent steep and gentle slopes.
4 Look carefully at the contour pattern in map B and sketches C. Explain which of the following labels 1–4 are showing: a valley; a steep slope; a ridge; a gentler slope.
5 Compare map-flaps A, B, D and E. Look carefully at the contour patterns on each map, and decide which maps predominantly show gentle slopes, flat land or steep slopes.
6 Look carefully at map-flap A, Nant Ffrancon, 1:25 000 OS map.
   a) Give the four-figure grid reference for an area you think has: a steep slope, a gentle slope, flat land.
   b) Explain your choice and the evidence you used.
   c) Give the six-figure grid reference for a spot height.
7 Look carefully at map-flap D, Swanage, 1:50 000 OS map.
   a) What kind of slope is located at 025812?
   b) What happens to this slope at 029818?
   c) What feature is located at 968757? What is the height of the land there?

Learning objectives

▶ To compare an atlas map with an OS map and an aerial photo.

▶ To compare a vertical aerial photo with an OS map at the same scale.

▶ To identify features and land uses on a map and an aerial photo.

Studying a selection of maps and images when carrying out enquiries enables you to have a fuller understanding of the location of a place. When investigating an OS map extract, it is useful to be able to compare it with an aerial photo and atlas map of the same area:

● **Aerial photographs** show what the land looks like from above. They are taken from aeroplanes or, increasingly, from satellites. It is possible to access aerial photographs of most places in the world using software such as Google Earth.

● **Atlas maps** are at a smaller scale than OS maps so are useful in identifying the location of places within the UK.

Photo A is a vertical photograph of Liverpool, showing the same area as map-flap B. Using OS maps alongside photographs like this allows you to identify both physical and human geographical features as well as different land uses, such as buildings, roads, rivers and industry.

Activities

1 Using your school atlas:
   a) Turn to the index and find Liverpool.
   b) Record the latitude and longitude for Liverpool.
   c) Find Liverpool on the page referenced in the index.
   d) Compare the atlas map with map-flap B. Name the features you can see on the OS map but that are not named – the river and the city to the west of Liverpool.
   e) Describe the absolute location of Liverpool using coordinates and grid references. (Look back at Lesson 1.4 to remind yourself how to describe location.)
   f) Describe the relative location of Liverpool using the atlas and OS maps.
2 What is a vertical aerial photograph?
3 Ten features are labelled with letters on photo A.
   a) Find these features on the photo and map-flap B. You will also find it useful to use the 1:50 000 OS symbols at the back of this book.
   b) Copy and complete the table to record the name and grid reference for features A–J.

Feature	Name of feature	Six-figure grid reference
Water area A		
Land use B		
Land use C		
Building D		
Building E		
Type of road F		
Water feature G		
Land use H		
Land use I		
Line of communication J		

4 You can find similar maps and aerial photos for your local area.
   Use either your school access to OS Digimap for schools, Google Earth or copies of photos and OS maps that your school may have of your local area.
   Create your own labelled map and photo for your local area showing different land uses, and important physical and human geography features.

A Vertical aerial photograph of Liverpool from Getmapping

Learning objectives

▶ To use basic calculations – mean and range.

▶ To be able to draw a line graph and bar chart.

▶ To interpret statistics about Liverpool using enquiry questions.

In Lesson 1.2, you were introduced to the term 'quantitative data'. Our twenty-first-century world is data rich; numbers and statistics are now an essential feature of life. A good geographer can understand and interpret this data. This lesson provides an introduction to using statistics and graphs to understand places.

Statistics

Statistics are all around us, on TV news, in newspapers, adverts, even on food packaging. Statistics are used to investigate new ideas or world events, and change. Exploring the use of statistical techniques helps geographers to analyse numerical data. Handling data properly can show trends and patterns, and help people reach conclusions about enquiry questions. Statistics only make sense if we can see a pattern or a trend in the data. This is why statistics are often converted into averages or graphs to make them easier to understand.

Working out some basic calculations from data

The mean is calculated by adding up the total numbers and dividing by how many numbers there are.

E.g. 4 + 12 + 12 + 16 + 17 + 12 + 4 = 77

The mean is 77 divided by 7 (the number of values) = 11.

The range is the difference between the largest and smallest values in a set of data. In the data set example above, 17 − 4 = 13.

Graphs

As a geographer, you should also be able to construct and interpret graphs from statistics. You will be provided with opportunities to do this throughout this book. Line graphs are used to show change. Bar charts compare different quantities.

Step 5: Add a title.

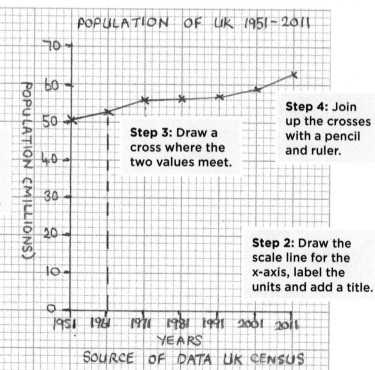

Step 1: Draw the scale line for the y-axis, label the units and add a title.

Step 3: Draw a cross where the two values meet.

Step 4: Join up the crosses with a pencil and ruler.

Step 2: Draw the scale line for the x-axis, label the units and add a title.

Step 6: Identify the source of the statistical data.

A How to draw a simple line graph

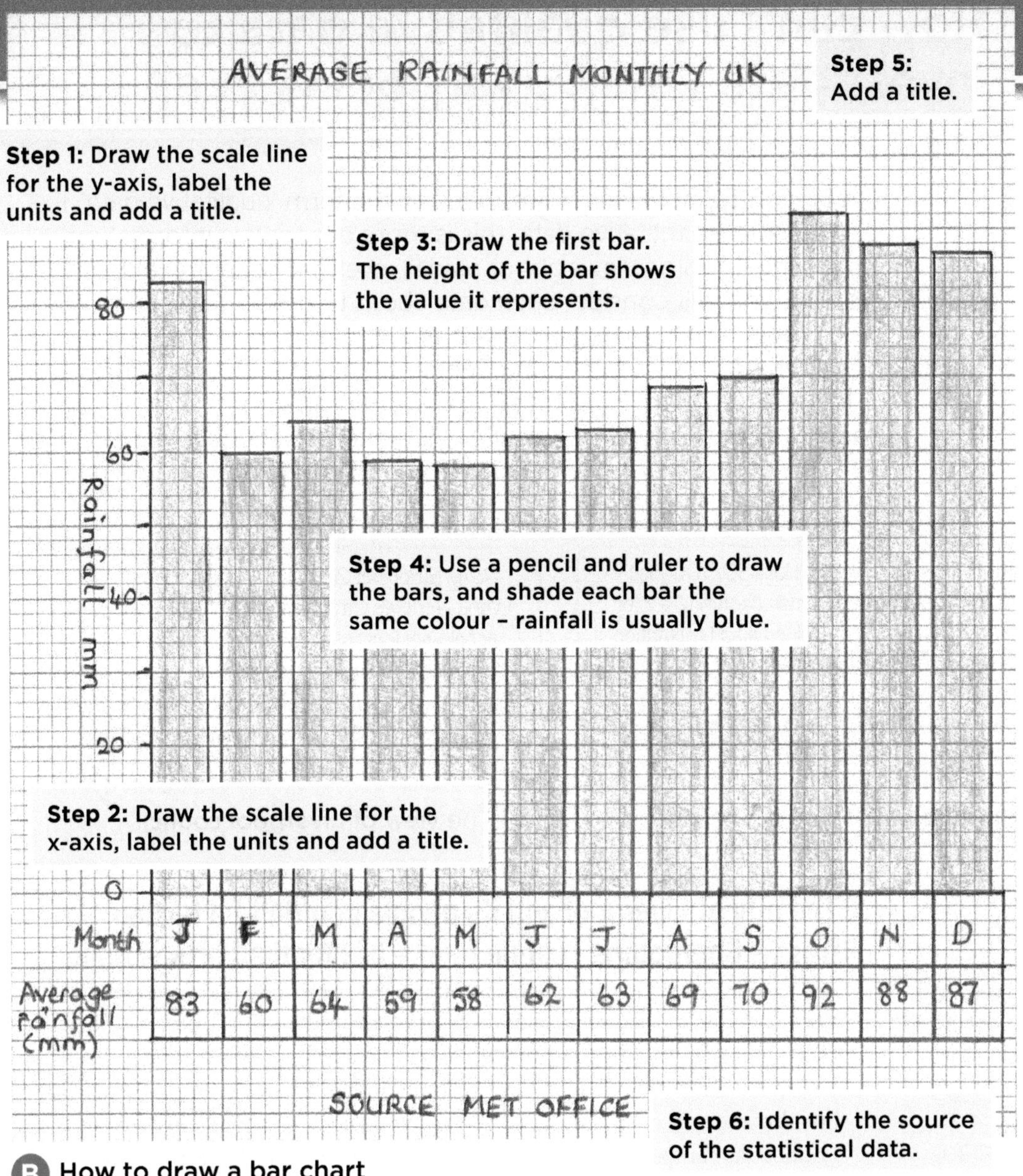

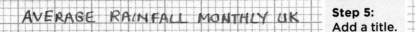

AVERAGE RAINFALL MONTHLY UK

Step 5: Add a title.

Step 1: Draw the scale line for the y-axis, label the units and add a title.

Step 3: Draw the first bar. The height of the bar shows the value it represents.

Step 4: Use a pencil and ruler to draw the bars, and shade each bar the same colour – rainfall is usually blue.

Step 2: Draw the scale line for the x-axis, label the units and add a title.

Rainfall mm

80
60
40
20
0

Month	J	F	M	A	M	J	J	A	S	O	N	D
Average rainfall (mm)	83	60	64	59	58	62	63	69	70	92	88	87

SOURCE MET OFFICE

Step 6: Identify the source of the statistical data.

B How to draw a bar chart

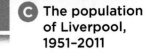

C The population of Liverpool, 1951–2011

Year	Population
1951	765,641
1961	737,637
1971	595,252
1981	492,164
1991	448,629
2001	439,428
2011	466,415

D The average monthly rainfall for Liverpool

Month	Rainfall (mm)
J	75
F	54
M	64
A	54
M	55
J	66
J	59
A	69
S	72
O	97
N	83
D	89

Activities

1 What is quantitative data?
2 How do geographers make statistics easier to interpret?
3 **a)** What is the mean?
   b) Calculate the mean rainfall for the UK, using the data in graph B.
4 **a)** What is the range?
   b) Calculate the range of rainfall for the UK.
5 Read the instructions in A.
   a) Use the instructions to draw a line graph to show population change in Liverpool, using the data in table C.
   b) Write three sentences to describe what the graph shows about how Liverpool's population has changed.

6 Read the instructions in B.
   a) Use these instructions to draw a bar chart to show the average monthly rainfall for Liverpool, using the data in table D.
   b) Calculate the total annual rainfall for Liverpool and write the figure beside your bar chart.
   c) Calculate the total annual rainfall for the UK as a whole using the data in graph B. How does it compare with the total annual rainfall for Liverpool?
   d) Compare graph B with your bar chart. Write four sentences to describe how the monthly rainfall patterns for the UK and Liverpool are similar or different.

Learning objectives

▶ To interpret qualitative data about Liverpool using enquiry questions.

▶ To consider people's different viewpoints about a place.

▶ To be able to compare a pictorial tourist map with an OS map.

In Lesson 1.2, you were introduced to the term 'qualitative data'. In contrast to quantitative data, this is more descriptive information, involving people's views about places, events or change. In this lesson, you will consider such data about Liverpool, including a pictorial tourist map, photographs, video and points of view. This type of data is richer and deeper than just facts and figures. However, it is important that you consider the origin, reliability and accuracy of this type of data. People's interpretations of a place may be biased, distorting the data to promote a particular viewpoint.

Once the great port of the British Empire, Liverpool lost 80,000 jobs between 1972 and 1982 as the docks closed and its manufacturing sector shrank by 50 per cent. As a consequence, the city gained a bad reputation due to high unemployment and economic decline. This resulted in mass poverty across the city and led many to view Liverpool as an undesirable place to live. Speak to someone who hasn't been to Liverpool before, the majority would say it's not a great place or that it's tatty.

A The views of a Liverpool resident

Being European Capital of Culture in 2008 and a World Heritage City has redefined Liverpool to many people, not least our residents and visitors, but also businesses, the media and opinion formers. The year gave the city a priceless platform to showcase a new, dynamic and creative Liverpool on a global stage.

B The view of Liverpool council

Inscribed a UNESCO World Heritage Site in 2004, Liverpool Maritime Mercantile City is part of a special family of World Heritage Sites that include the Taj Mahal, Stonehenge, Venice and the Great Wall of China. Liverpool developed into one of the world's major trading ports in the eighteenth and nineteenth centuries. The city played an important role in the growth of the British Empire.

C UNESCO reasons for Heritage City status

The city of Liverpool and its people couldn't be further from a Champions League final – but football brings hope

Eleven of Liverpool's neighbourhoods are among the most deprived one per cent in the country. Life expectancy is six and a half years lower than the national average. Foodbanks have become both a symbol of the desperation but also a source of salvation and even hope. The Fans Supporting Foodbanks campaign, which involves both Liverpool and Everton, now supplies around 20 per cent of all foodbank donations in Liverpool.

D *The Independent*, 22 May 2018

Activities

1 What is qualitative data?

2 The five examples of data about Liverpool, A–E, present different views about the city.
   a) Sort the data into two lists – positive and negative views of the city. Some of the data include both negative and positive views, so will appear in both lists.
   b) In each case, give a reason or evidence for your choice.

3 Read view A. Describe how Liverpool changed in the 1970s, and how this affected its reputation.

4 Read B and C. Describe the developments that have led to Liverpool improving.

5 Look carefully at D. Describe how this data presents a different view of the city.

6 Watch the online video 'Liverpool: A day in the life' with a partner: https://vimeo.com/68071838
   a) Discuss with your partner different words that best describe the Liverpool portrayed in the video.
   b) Compare the video's image of Liverpool with data A–E. Write four sentences to identify similarities and differences.

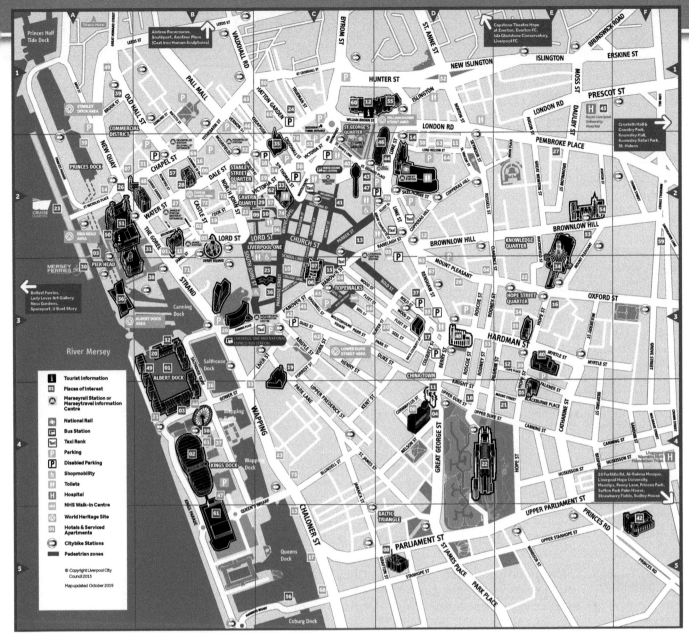

🅔 **Tourist map of Liverpool,** http://visitliverpool.com

7 Look at map E, which is a pictorial map of Liverpool.
  a) Explain what sort of view or image of the city the map is presenting.
  b) Compare map E with the OS map-flap B, of Liverpool.
    Map E uses grid references but they are different from OS grid references in that they use letters and numbers. Locate which grid squares on map-flap B the area shown in map E is in.
    Locate the following on the two maps and record their OS grid references and coordinates on the visitor map: Liverpool Lime Street railway station; Albert Dock; No. 38 Mersey Ferry wharf; the road junction of Parliament Street and St James Place.

8 OS maps and visitor maps like E have been designed for different purposes and usage. Identify three ways people would use visitor map E that are different from how they would use OS map-flap B.

9 Write five sentences to summarise five key things you have discovered about what Liverpool is like.

Learning objectives

▶ To understand what GIS is.

▶ To know ways GIS can be used to investigate places.

▶ To become familiar with GIS packages.

Access to geographical data is being revolutionised with the development of Geographical Information Systems, or GIS.

What is GIS?

GIS uses computers and software designed to capture, store, manipulate, analyse, manage and present data. This data can be tied to a particular location on the ground, such as a postcode, latitude and longitude, building, or road junction. The GIS lets you visualise map data. GIS works with different layers of information about the same geographical area at the same time. These layers of data can be linked to a range of different base maps and satellite imagery (see diagram A). When using the GIS application, the user can switch these layers on and off, depending on the enquiry question they are investigating. The GIS will only provide useful answers to problems if the user is able to ask the right geographical questions and can interpret the results.

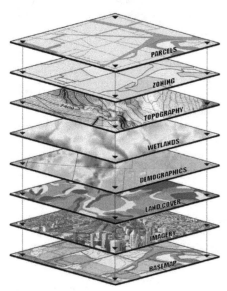

A GIS layers

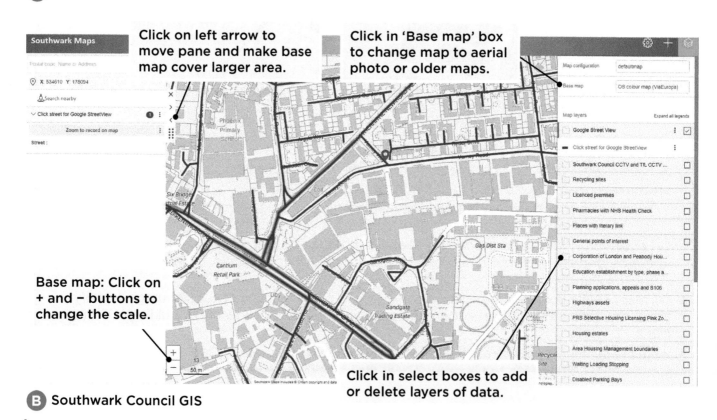

Click on left arrow to move pane and make base map cover larger area.

Click in 'Base map' box to change map to aerial photo or older maps.

Base map: Click on + and − buttons to change the scale.

Click in select boxes to add or delete layers of data.

B Southwark Council GIS

Who uses GIS?

There is a large industry in GIS software with hundreds of companies producing thousands of products. Ordnance Survey, Google and ESRI are examples. GIS is used by individual people, communities, research institutions, environmental scientists, health organisations, land use planners, businesses and government agencies at all levels. Many government agencies use GIS software to help in planning and organising their geographic data. Southwark Council in London, for example, has created a GIS on its website (see screenshot B).

What is Google Earth?

The giant USA-based technology company Google has transformed the way we see the world. They launched Google Earth in 2005. Since then, they have developed other spatial technology, including Google Street View. These are all online tools that are free and relatively easy to use. Google Earth represents the Earth as a three-dimensional globe. It allows users to get a bird's-eye view of the planet and then zoom into particular locations. As you search locations around the world, the latitude and longitude is provided in the corner of your screen.

ESRI

The Environmental Systems Research Institute (ESRI) is an international supplier of GIS software and web GIS. The company headquarters is in California, USA. ESRI is a global company, with 49 offices worldwide and employees from 73 countries. The company's GIS system is called ArcGIS.

Activities

1 What is GIS?
2 How is data organised in a GIS?
3 How does GIS software connect different data to a base map?
4 On the internet, use a search engine to search for 'Southwark Council Map Service'. Launch the maps.
   a) Identify examples of different layers of data on the site.
   b) Click in the base map box. You will see that there is a wide range of maps here. Try out a few.
   c) Imagine you are a local resident of the London Borough of Southwark. Identify three ways you might use this GIS.
5 Go to the following website: www.google.com/earth/education. On this website, you can find out how to use Google Earth. Launch Google Earth and search for Liverpool. Explore the city and identify five key points you learnt that add to what you have already learnt about Liverpool.
6 ESRI has created a learning area on their website to help schools understand how to use the software. Use an online search engine and search ArcGiS online for schools. Visit the website and discover how you can use the software in your geographical studies. Identify five ways you can use ArcGiS.
7 Look at the vision statement, map-flap C, and think about what you have learnt about GIS.
   Write a paragraph to explain how you now think using GIS will develop your skills as a geographer.

Future learning

You will have opportunities to further progress your capabilities using GIS in other units of this book.

Learning objective

▶ To know how to draw a fieldsketch.

Fieldsketching is an important geographical skill when conducting fieldwork. Identifying and recording key geographical features you can see on the ground, in a fieldsketch, helps you to interpret landscapes as a geographer. It makes you carefully observe a landscape, comparing what you can see to an OS map. Fieldsketches are best drawn from a viewpoint, in the stages shown in B. This fieldsketch shows Scarborough in North Yorkshire. You can find Scarborough and the key features drawn on the fieldsketch on the OS map, A.

Activities

1 Use the fieldsketch B and OS map A to write a complete title for this fieldsketch:

A fieldsketch of Scarborough from the viewpoint at _____ Mount, six-figure grid reference _____, looking in a _____ direction.

2 Compare the final fieldsketch with OS map A.

Make a copy of the following table. Name the features labelled A–M on the final sketch and, where possible, record the six-figure grid reference. In each case, a clue is provided.

	Clue	Feature	Six-figure grid reference
A	Leisure activity		
B	Form of transport		
C	Bare rock on beach		
D	Name of coastal area		
E	Terminus for railway		
F	Class of road		
G	Place of worship		
H	Coastal feature		
I	Building for a boat		
J	Historical building		
K	Building		
L	Harbours		
M	Name of coastal area		

3 Draw a fieldsketch of your local area, or your school grounds and surrounding area.
 – Look for a viewpoint that provides a good view.
 – Draw your sketch in the stages shown in B.
 – Use an OS map and your local knowledge to label geographical features, as shown in B.

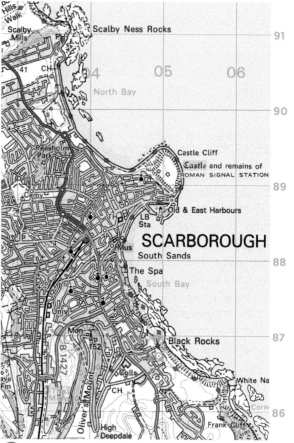

A   1:50 000 OS map of Scarborough

© Crown copyright and database rights 2020 Hodder Education under licence to Ordnance Survey

Equipment for fieldsketching

● Plain sheet of A4 paper, landscape view
● Pen, pencil, coloured pencils, eraser
● Clipboard
● OS map of your locality, either 1:25 000 or 1:50 000

B How to draw a fieldsketch

Step 1: Locate a viewpoint (in this case the viewpoint at Oliver's Mount). Look carefully at the view and identify geographical features with an OS map.

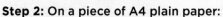

Step 2: On a piece of A4 plain paper:
- draw a horizon line two-thirds up the page
- decide on the limits of your sketch
- draw the key features in the correct places to provide a rough outline of your sketch.

Step 3: Once you are happy with your outline, add more detail to your fieldsketch.

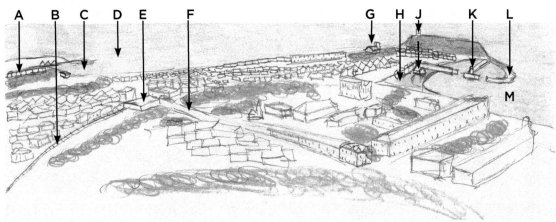

Step 4: Add colour to your fieldsketch to help make geographical features stand out. Use the OS map to help you label the key features – draw vertical arrows from each feature into the top third of your sketch, which is for the labelling.

Learning objectives

▶ To locate photographs on OS maps of different scales.

▶ To compare ground-level photos with an OS map.

▶ To identify geographical features on a map and photo.

A good geographer investigates places by conducting fieldwork. When visiting a location, you can use maps and observation to collect, record and present data. A group of students conducted fieldwork at Swanage, a seaside resort on the Dorset coast. They used OS maps of different scales, 1:50 000 map-flap D and 1:10 000 map A, below. They marked locations along the route they walked on map A, 1–7, and took photographs to record physical and human features they saw, B–H.

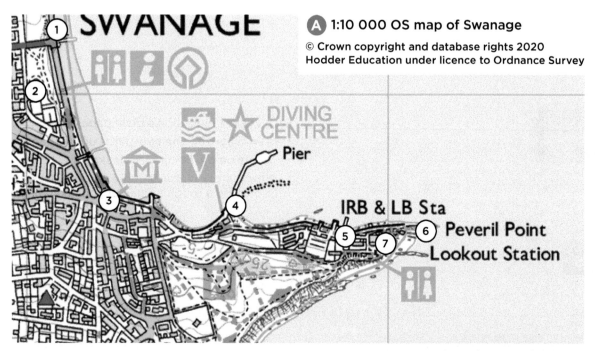

A 1:10 000 OS map of Swanage

© Crown copyright and database rights 2020
Hodder Education under licence to Ordnance Survey

E

F

H

1

2

4

3

Activities

1 Compare map A and map-flap D.
  a) Identify the route shown on map A, on map-flap D.
  b) Measure the distance the students walked along the route.
  c) In which direction did they walk?
2 Look carefully at photos B–H and match them with the locations marked 1–7 on map A. There are clues on each photo which link to OS symbols and labels on the map. In each case, give evidence from the photo and map to justify your choices.
3 Compare map A with map-flap D. Give the six-figure grid reference for each location 1–7 on map A.
4 Write a description of the route the students followed, using the OS maps and photos, adding the name and six-figure grid reference of each location along the route.
5 Photo H was taken at 030793.
  a) Locate this grid reference on map-flap D. In which direction was the camera pointing?
  b) Draw a fieldsketch of the view, using the guidance in Lesson 1.13.
  c) Label and name features 1–4 on your fieldsketch.
  d) Add a title to your fieldsketch.
6 You could conduct similar fieldwork for the locality around your school.

Future learning

This unit has introduced geographical enquiry as well as a range of geographical skills, including atlas and OS maps, statistics, graphs, qualitative data, photographs, GIS and fieldwork.

You will be provided with opportunities to develop and progress these skills throughout *Progress in Geography Skills*.

Learning objectives

▶ To compare atlas maps of Britain showing physical geography and geology.

▶ To know the main physical regions of the UK.

▶ To identify the relationship between physical landforms and geology.

▶ To identify rocks used as a natural resource.

Map B is a physical atlas map of Britain, showing the relief features of the country. Relief is a word used to describe the physical features of the land. Britain is fairly small, and yet it has a wide variety of relief features – mountains, valleys, lowlands. This variety is the result of a complicated geology – the rocks that form the landscape (shown in map A). Hard, resistant rocks, such as granite and slate, form the rugged mountain ranges of Scotland. Weaker rocks, such as clay, often form low-lying plains. Different rocks are used by people as a natural resource, for energy, and as a building material.

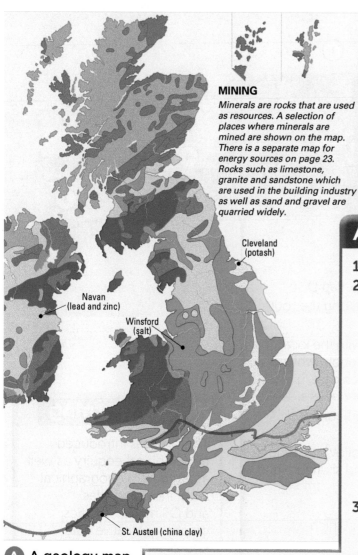

MINING

Minerals are rocks that are used as resources. A selection of places where minerals are mined are shown on the map. There is a separate map for energy sources on page 23. Rocks such as limestone, granite and sandstone which are used in the building industry as well as sand and gravel are quarried widely.

Cleveland (potash)

Navan (lead and zinc)

Winsford (salt)

St. Austell (china clay)

Rock type	Geological Era
Sands and clays	TERTIARY (0–65 million years old)
Chalk	
Clays, sands, sandstone	SECONDARY (65–230 million years old)
Limestone	
Coal measures	
Limestone, millstone grit	PRIMARY (230–570 million years old)
Sandstone	
Shales and slates	
Gneiss, quartzite, schists	Various ages
Basalt and granite	

A A geology map of the United Kingdom, © *Philip's Essential Atlas*

Activities

1 Write a definition of geology.

2 Look carefully at map B.

   a) Compare the map with the key. What is the height of the upland and lowland areas?

   b) Name the highest mountains in Scotland, Wales and England. For each, write down:

      i) its height above sea level

      ii) which mountain range it is in.

   c) Name the range of mountains that runs north to south in the middle of England.

   d) Write a paragraph to describe the distribution of mountain and lowland areas in the UK.

3 Compare maps A and B. Copy and complete the following table to identify the rocks that each relief feature is formed from.

Relief feature	Geology	Relief feature	Geology
Cambrian Mountains		Dartmoor	
Lake District		Pennines	
North York Moors		North-West Highlands	
North and South Downs		London – Thames Valley	

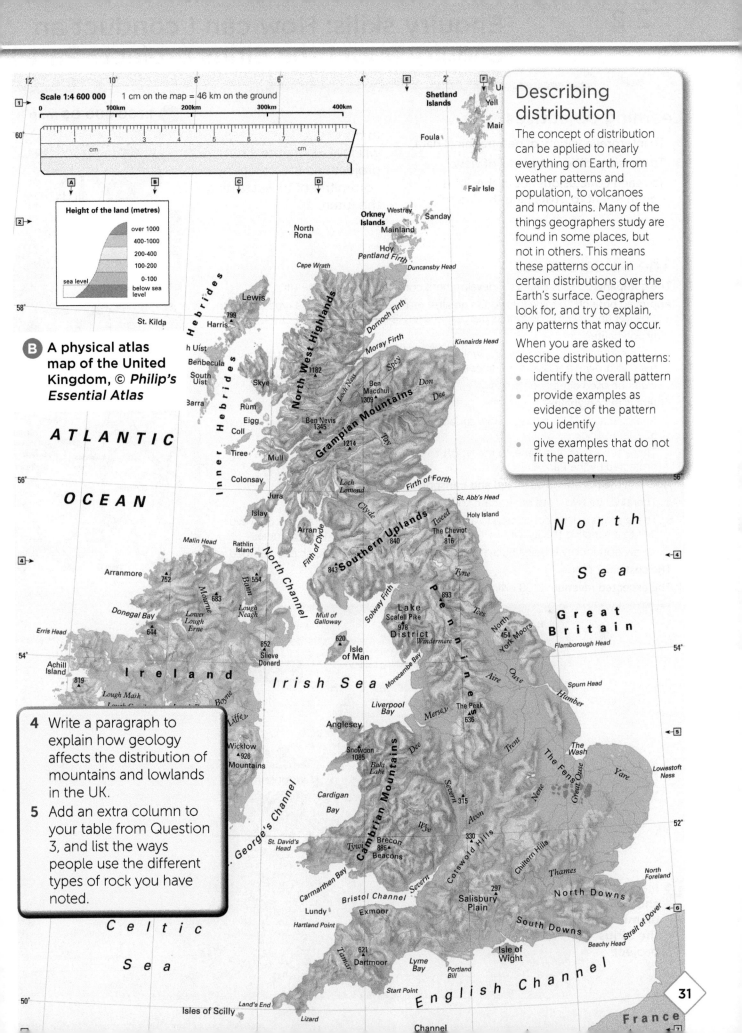

Describing distribution

The concept of distribution can be applied to nearly everything on Earth, from weather patterns and population, to volcanoes and mountains. Many of the things geographers study are found in some places, but not in others. This means these patterns occur in certain distributions over the Earth's surface. Geographers look for, and try to explain, any patterns that may occur.

When you are asked to describe distribution patterns:

- identify the overall pattern
- provide examples as evidence of the pattern you identify
- give examples that do not fit the pattern.

Scale 1:4 600 000   1 cm on the map = 46 km on the ground

Height of the land (metres)

- over 1000
- 400–1000
- 200–400
- 100–200
- 0–100
- sea level
- below sea level

B **A physical atlas map of the United Kingdom,** © *Philip's Essential Atlas*

4 Write a paragraph to explain how geology affects the distribution of mountains and lowlands in the UK.

5 Add an extra column to your table from Question 3, and list the ways people use the different types of rock you have noted.

31

France

2.2

Enquiry skills: How can I conduct an enquiry about the Sirius project? Part 1

Learning objectives

▶ To interpret a variety of geographical data.

▶ To consider different points of view.

▶ To reach conclusions about the Sirius commitment to 'sustaining the future'.

In these two lessons, you will investigate the Sirius project and the company's commitment to 'sustaining the future'.

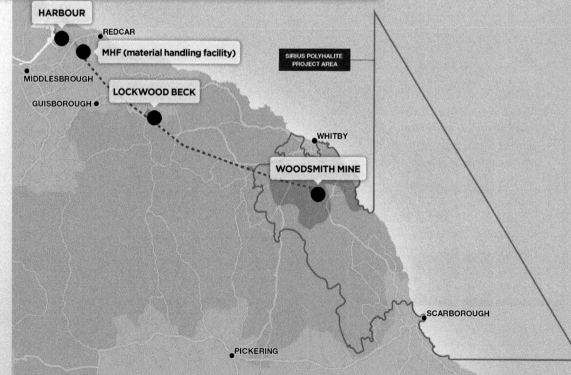

A 1:200 000 OS map

The Sirius project

The company: Sirius is a fertiliser development company, based in the UK.

Approved project: To mine the world's greatest reserves of polyhalite – a type of potash – to deliver up to 20 million tonnes per year of a fertiliser product called POLY4.

Why polyhalite?

- It is a naturally occurring mineral which contains four of the six nutrients plants need to grow – potassium, magnesium, sulphur and calcium.
- It can be used directly on crops.

Where is the polyhalite?

- Approximately 1550 metres below surface level, in deep underground seams up to 50 metres thick.
- These seams are also beneath the North York Moors National Park – an area protected by the Park Authority.

How will the polyhalite be mined and transported?

- There will be two deep mine shafts to access the polyhalite.
- These shafts will be connected to a tunnel to transport the mineral on a conveyor belt to a handling facility, where it will be processed and turned into granules.
- A new port facility is being built at Teesport, from where it will be shipped abroad.

The cost: £3.2 billion.

The expected revenue: £100 billion over the next 50 years.

B Map showing elements of the Sirius project

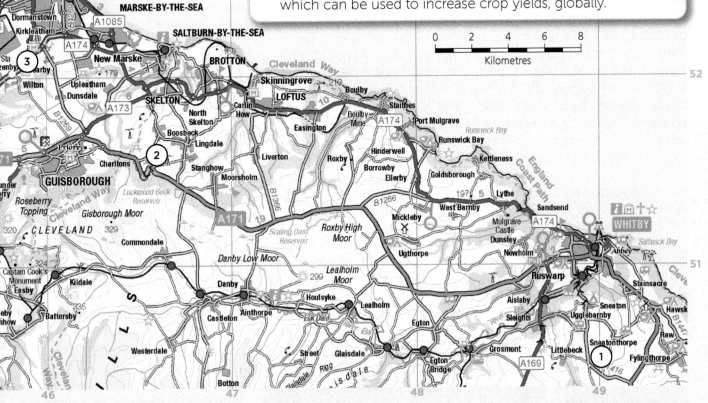

© Crown copyright and database rights 2020 Hodder Education under licence to Ordnance Survey

Commitment to sustainable development

Sirius is committed to minimising the project's impact on the environment, by:

- seeking planning permission from the Park Authority – it took four years to get permission to begin the mining
- keeping visual impact to a minimum
- supporting the global demand for food, for an increasing world population – POLY4 is a naturally occurring fertiliser which can be used to increase crop yields, globally.

C **Cross-section of the Sirius development**

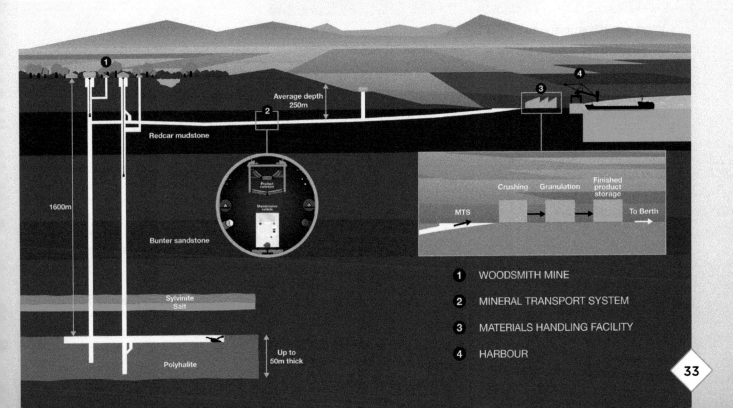

1 WOODSMITH MINE

2 MINERAL TRANSPORT SYSTEM

3 MATERIALS HANDLING FACILITY

4 HARBOUR

2.3

Enquiry skills: How can I conduct an enquiry about the Sirius project? Part 2

In June 2015 the North York Moors National Park Authority gave planning permission for the development of the mine. A shaft a mile deep will be dug underneath the protected moorland, as well as a 16km tunnel to carry the mineral from the mine to the Materials Handling Facility, where it will be crushed before loading onto ships at a new terminal at Teesport. The vote to give Sirius mining permission was close, eight voted in favour, with seven against. The project was supported by 93% of local people who wrote to the council, 81% of whom live in the national park. But it was opposed by groups including the National Trust, Natural England, Campaign for National Parks, who described it as a "huge threat" to the moors. It took four years for the company to gain permission, they worked hard to demonstrate how the project would be sustainable, as well as bringing jobs and money into a neglected part of the country. Photos D shows how the mine will be disguised. Different viewpoints about the project are shown in views E to J. Sirius describe polyhalite as a "fertiliser of the future". The firm estimates there is enough polyhalite to keep the mine open for at least 100 years, providing more than 1,000 jobs and generate £1bn a year in exports. 300 farmers and other landowners have struck lucrative royalty deals for the mineral rights under their land. The company expects to pay out £1.4 billion in royalties in the first 50 years of operation.

Before | During | After

D Woodsmith mine visualised by Sirius

> We raised concerns over the Woodsmith Mine when it was first proposed in 2013. It's really disappointing that their plans continue to present challenges to reducing the mine's impact on the North York Moors.

G Ruth Bradshaw, Policy and Research Manager, Campaign for National Parks

> Building a mine of this scale in a National Park is simply at odds with the whole purpose of National Parks. The changes that are now being proposed will degrade the special qualities of the National Park still further and we are beginning to see the impact of the development at the minehead site at Woodsmith, what was Doves Nest Farm.

H Tom Chadwick of the North Yorkshire Moors Association

> In delivering this project, we can create thousands of jobs in North Yorkshire and Teesside, deliver billions of pounds of investment to the UK and put the country at the forefront of the multi-nutrient fertiliser industry.

I Chris Fraser, the Managing Director and Chief Executive of Sirius

> Over the next 35 years, the Earth is being asked to produce more food than it has to date in human history ... The North York Moors is an area of significant international and national value, and as an ecologist, I value such areas highly. Whilst I do not under-estimate the impact of the construction work, this should be offset against the potential importance at a global level of the polyhalite deposits. This includes enhancing global sustainability by improving yields.

E Professor Tim Benton, Head of UK's Global Food Security Programme

> It will be a great boost for the local economy and should not be opposed simply because it is 'in our own back yard'. Please approve the application and stop the deteriorating living standards of the Whitby people. Stop the young moving out and plan to keep families together.

F Mr K Froggatt from Moorside Farm in Littlebeck

> A wide range of different matters, such as impact on the landscape, wildlife and local residents, need to be balanced against the benefits of the mine – for example, new jobs for people, and supply of a new source of fertiliser. In this particular case, the Authority took the view that the economic benefits of the mine were so important that the development would result in a wider public benefit.

J Rob Smith, Senior Minerals Planner, North York Moors National Park Authority

Activities

1. Look through all the geographical data and text about the Sirius mining project for part 1 and part 2 of this enquiry lesson. Answer the introductory questions.
   a) When did Sirius gain planning permission to develop the mine?
   b) How long did it take the company to gain planning permission?
   c) Why did Sirius need to get planning permission for this project?
   d) How did the members of the planning committee vote?
   e) Why was the decision so close?
   f) Why do you think local farmers would be keen for the development to go ahead?

2. Compare the OS map A with the Sirius map B.
   a) Match locations 1–4 on map A with the following key elements of the project on map B: mine at Woodsmith (near the village of Sneatonthorpe); the second shaft at Lockwood Beck; the Material Handling Facility; the port at Teeside.
   b) Use map A to give the six-figure grid reference for each element.
   c) Initially, Sirius may have thought about transporting the mineral from the mine to the port by road. Identify the best route on the OS map and measure the distance along the roads from the mine to the port.
   d) Sirius has now built a tunnel. Measure the distance of the route of the tunnel shown on map B, on map A.
   e) Write three sentences to explain why Sirius built a tunnel rather than using the roads.

3. Look carefully at map B and cross-section C.
   a) How deep are the deposits of polyhalite?
   b) Write a paragraph to explain how the four elements of the Sirius project work together, with minimum environmental impact on the National Park.

4. Look carefully at photos D and write a paragraph to explain how Sirius plan to protect the environment of the National Park, near the Woodsmith Mine.

5. Read the viewpoints about the project, E–J.
   a) Write two lists that identify the people for and against the scheme.
   b) In each case, give a reason for your choice.
   c) Identify what you think is the most important viewpoint, and explain your choice.

6. Use all the geographical data in these two lessons to create a list of the advantages and disadvantages of the Sirius project.

7. Conclusion: Write a paragraph to explain how you think Sirius Minerals is attempting to 'Sustain the Future'.

How can I draw pie charts to understand the economy?

Learning objectives

▶ To be able to draw and interpret pie charts.

▶ To understand that the jobs people do can be categorised into sectors.

The economy of a country includes all the different types of jobs that people do for a living. These jobs can be divided into four groups, called the employment sectors or structure of a country:

- **Primary**: involves the extraction and production of raw materials, from the land and sea. Sirius Minerals in Lessons 2.2 and 2.3 is in the primary sector, as are jobs relating to farming, forestry and fishing.

- **Secondary**: involves making things or manufacturing raw materials into goods, for example manufacturing steel into cars. All manufacturing, processing and construction jobs lie within this sector.

- **Tertiary**: provides a service to others. Jobs include teachers, doctors, refuse collectors, shop assistants and utility workers such as water, electricity and gas companies.

- **Quaternary**: in the late twentieth century, it began to be argued that traditional tertiary services could be further divided into a quaternary sector. In this sector, workers have a high level of expertise and skills, often earning high salaries. Jobs include scientific research, and the development and use of new technologies, as well as intellectual services such as financial support and advice. Some sources of data include quaternary jobs within the tertiary sector.

What is a pie chart?

Geographers often use pie charts to analyse the employment structure of a region or country, and to compare the structure in one region or country with another. Pie charts are useful tools for this because they show, at a glance, the different proportions of each sector. In this lesson, you are going to learn how to draw pie charts to compare the employment structure within the UK.

A pie chart is a circular graph which is divided up into segments. Each segment represents a proportion of the whole. A complete pie chart is 360° so when you are dealing with percentages each percentage point is equal to 3.6° on your pie chart $(\frac{360}{100})$.

Step 7: Add a title to show what the pie chart shows.

Step 1: Draw a circle with a pencil and compass. Then draw a line from the centre to 12 o'clock.

Step 6: Shade each segment a different colour. Add labels and % for each to show what they represent.

Step 5: Check the angle for the final segment, which should be 24 x 3.6 = 86°.

Step 2: The primary sector is 1%, which is 3.6°. Use a protractor to mark the angle for this and draw a second line from the centre of the circle.

Step 3: The secondary sector is 18% so 18 x 3.6 = 64.8°. Line the protractor along the second line you have drawn, and mark the angle, then draw another line from the centre.

Step 4: Repeat the process for the next sector, which is 57 x 3.6 = 205°.

Employment Structure for UK 2011
Primary 1%
Secondary 18%
Quaternary 24%
Tertiary 57%

A How to draw a pie chart of the UK economic sectors in 2011 (see table B)

Note for drawing pie charts showing employment structure: Normally the different segments of a pie chart would be given clockwise in size order, with the largest shown first. In this case, we are looking at the different employment sectors in order, so the segments are in the order primary, secondary, tertiary, quarternary.

Note for drawing pie charts to compare employment structures for different countries: If we were comparing the UK's employment structure with the structure in another country at the same time, we would use the same ordering and shading of our segments to make it easier to compare them.

B Economic sectors in the UK from 1791 to 2011

	1791	1841	1891	1991	2011
Primary	75%	22%	15%	3%	1%
Secondary	15%	51%	55%	28%	18%
Tertiary	10%	27%	30%	54%	57%
Quaternary	O	O	O	15%	24%

Activities

1 Write definitions for each of the sectors of the economy. Give two examples of jobs for each sector.
2 Write a sentence to explain what a pie chart is and what it shows.
3 Write a paragraph to describe what pie chart A is showing.
4 Draw pie charts to show the employment structure for the two countries shown in table C. Use the same colours for each segment as those used in A.

C Employment sectors in Niger and the UAE, 2011

Country	Primary	Secondary	Tertiary
Niger	42%	19%	39%
United Arab Emirates	0.9%	50%	49.1%

5 Compare your two pie charts with A.
  a) Explain why it was important that you use the same colours in the segments in your pie charts as A.
  b) Write a paragraph describing the similarities and differences in the employment structure of the UK, Niger and the UAE.
6 Study table B. Pie chart A has been drawn to present the data for 2011. To show changes in the employment structure over time, a graph is the best method. A pie chart is not an appropriate method for this.
  a) Draw a line graph to plot the data for each year. Draw the line graph for each sector on the same axis, but use a different coloured line to represent each sector.
  b) Write a paragraph to describe how the percentage share of each sector changed between 1791 and 2011.

How can I use geographical data to determine what the port of Liverpool is like?

Learning objectives

▶ To understand what a port is and how it functions.

▶ To compare a plan with an OS map and satellite images on Google Earth.

▶ To use a GIS to investigate the range of ships in a port.

A port is a location on the coast that provides facilities for ships to load and unload their cargo. As well as being located at a sheltered location, a port also needs good transport links to move goods across the country.

The port of Liverpool is the most important UK deep sea container port for services between the UK and North America. It is ranked fifth in the UK in terms of tonnage. It sits on both banks of the River Mersey. The port benefits from direct links to the M53, M57, M62 and M6 (M58) motorways. The port is owned by Peel Ports. Map B is the official map of the port. Each year the government's Department for Transport publishes statistics for each port in the UK. Table A shows the weight of four categories of cargo for Liverpool: liquid bulk – crude oil and gas; general cargo; roll-on, roll-off, e.g. lorries (ro-ro); lift-on lift-off containers (lo-lo). In this lesson, you will use a variety of geographical data to determine what the port of Liverpool is like.

A Port of Liverpool by tonnage of cargo, 2017, Department for Transport

Type of cargo	Total tonnage (1000)
Liquid bulk	10,768
General cargo	8,598
Ro-ro	7,751
Lo-lo	5,424
Total	32,541

A pie chart can be drawn from raw data like that shown in table A. You will need to convert the total tonnage for each type of cargo into a percentage using the following formula:

$$\frac{\text{Tonnage of type of cargo}}{\text{Total tonnage}} \times 100 = \text{percentage of cargo type}$$

For example: percentage of liquid bulk cargo is:

$$\frac{10768}{32541} \times 100 = 33\%$$

Activities

1 Write a definition of a port.

2 Identify three advantages Liverpool has as a port, using map B and map-flap B.

3 Look back at A in Lesson 3.1, and the table of statistics on this page, A.

   a) Draw a pie chart to show the data, using the formula to calculate the raw data as percentages.

   b) Which are the most important cargoes in terms of tonnage?

4 A port has a wide range of different dock facilities for different cargoes and ships. For example, Tranmere Oil Terminal, shown on map B, is where oil tankers off-load liquid bulk. Look carefully at map B, and identify five other types of facility.

5 Compare map B and map-flap B. Locate the following facilities shown on map B with a six-figure grid reference on map-flap B:

Liverpool Dock Estate: 3 Royal Seaforth Dock; 10 Alexandra Dock; 17 Canada Dock

Birkenhead Dock Estate: 6 Vittoria Dock; 7 East Float; 9 Alfred Dock

6 Go to the following website: www.marinetraffic.com/en/ais/details/ports/432. This website provides a GIS for the port of Liverpool.

   a) How many ships are currently in port?

   b) How many are expected to arrive today?

Port of LIVERPOOL
Official Map

PEELPORTS

B Port of Liverpool official map

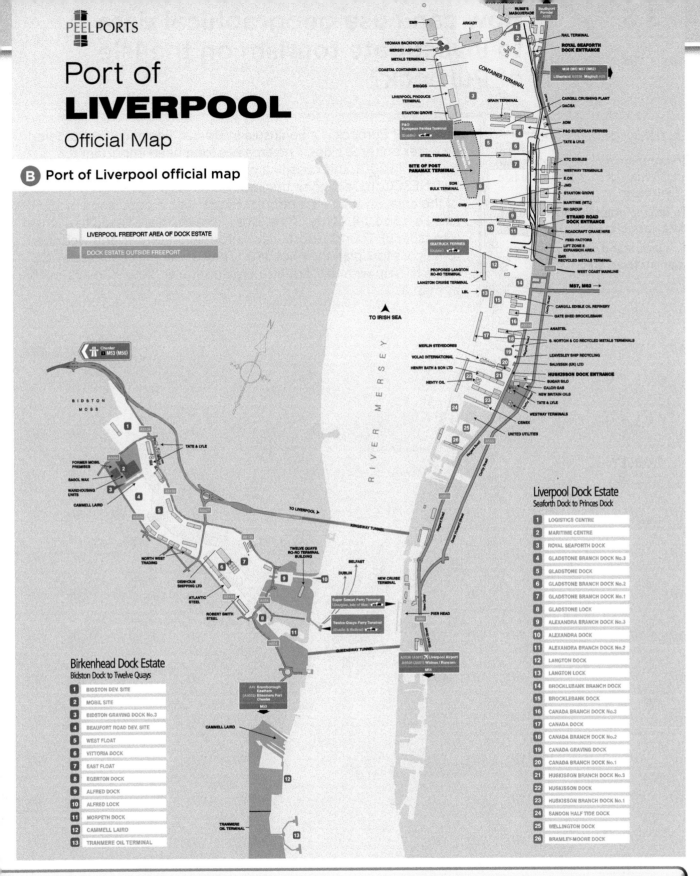

LIVERPOOL FREEPORT AREA OF DOCK ESTATE

DOCK ESTATE OUTSIDE FREEPORT

Liverpool Dock Estate
Seaforth Dock to Princes Dock

1	LOGISTICS CENTRE
2	MARITIME CENTRE
3	ROYAL SEAFORTH DOCK
4	GLADSTONE BRANCH DOCK No.3
5	GLADSTONE DOCK
6	GLADSTONE BRANCH DOCK No.2
7	GLADSTONE BRANCH DOCK No.1
8	GLADSTONE LOCK
9	ALEXANDRA BRANCH DOCK No.3
10	ALEXANDRA DOCK
11	ALEXANDRA BRANCH DOCK No.2
12	LANGTON DOCK
13	LANGTON LOCK
14	BROCKLEBANK BRANCH DOCK
15	BROCKLEBANK DOCK
16	CANADA BRANCH DOCK No.3
17	CANADA DOCK
18	CANADA BRANCH DOCK No.2
19	CANADA GRAVING DOCK
20	CANADA BRANCH DOCK No.1
21	HUSKISSON BRANCH DOCK No.3
22	HUSKISSON DOCK
23	HUSKISSON BRANCH DOCK No.1
24	SANDON HALF TIDE DOCK
25	WELLINGTON DOCK
26	BRAMLEY-MOORE DOCK

Birkenhead Dock Estate
Bidston Dock to Twelve Quays

1	BIDSTON DEV. SITE
2	MOBIL SITE
3	BIDSTON GRAVING DOCK No.3
4	BEAUFORT ROAD DEV. SITE
5	WEST FLOAT
6	VITTORIA DOCK
7	EAST FLOAT
8	EGERTON DOCK
9	ALFRED DOCK
10	ALFRED LOCK
11	MORPETH DOCK
12	CAMMELL LAIRD
13	TRANMERE OIL TERMINAL

c) Find the container terminal on map B and, on the website, click on a ship at the terminal. Find out and record the name of the ship, where it sailed from, and its cargo.

7 Use what you have discovered in this lesson to write a paragraph to describe what the port of Liverpool is like.

How can I use geographical data to investigate tourism on the Isle of Purbeck?

Learning objectives

▶ To compare a pictorial map with an OS map of tourism in the Isle of Purbeck.

▶ To draw a graph using statistical data about tourism in Swanage.

▶ To understand the importance of tourism to an area.

The Isle of Purbeck is a peninsula in Dorset, England. It is bordered by water on three sides. The area has long been important for tourism. In 2001, this coastline was awarded World Heritage status by UNESCO, to recognise the importance of the physical geography of the area. Swanage is a popular resort, with a population of 10,600 (see Lesson 1.14). Many of the jobs in the town are linked to tourism (see table B). Tourist resorts have information offices and websites to help visitors plan their holidays. They usually sell and display OS maps to help with this, but also create simpler pictorial maps, such as map A.

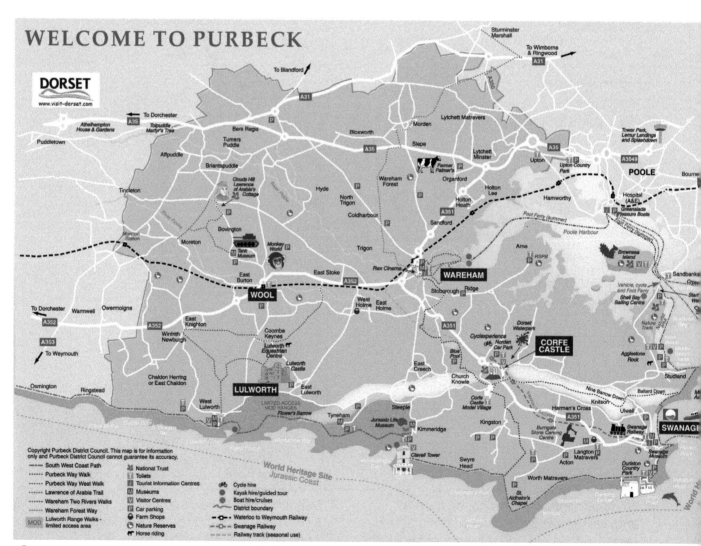

A Tourist map of the Purbeck region, **www.visitswanageandpurbeck.co.uk. Contains OS data** © Crown Copyright LA0100060963 2018 © Dorset Council

B Swanage tourism employment, Dorset Council

Categories of jobs related to tourism	Numbers
Accommodation	346
Retailing	135
Catering	445
Attractions	198
Transport	69
Total	1,193

C Two tourist attractions on the Isle of Purbeck

Activities

1. Use a road atlas or an online route planner such as Google Maps to plan a journey from your home to Swanage.
   a) How far is the journey?
   b) How long would it take?
   c) Describe the route you would take.

2. Go to the UNESCO World Heritage website: https://whc.unesco.org/en/list/
   a) Enter the search terms: 'Dorset and East Devon coast'.
   b) Identify why UNESCO awarded the coast World Heritage status.
   c) Look carefully at map-flap D and give the six-figure grid reference for the World Heritage symbol at Swanage.

3. Look carefully at table B. It shows that tourism is important for employment in Swanage.
   a) Calculate the employment numbers into percentages of the total figure.
   b) Think about what you have learnt about graphs. Decide which graph type you should use to present this data, and draw it.
   c) Write three sentences to describe what your graph shows.
   d) Look back at pie chart A in Lesson 3.1. Which employment sector dominates employment in Swanage?

4. Photo C was taken at the following grid reference: 963821.
   a) Find the location on map-flap D and map A. Name the two important tourist attractions shown in the photo.
   b) In which direction was the camera pointing when the photo was taken?

5. Look carefully at map-flap D. Identify ten different pieces of map evidence that the Isle of Purbeck area is important for tourism. Record your evidence in a copy of the table below.

Place name and evidence	Drawing of the map symbol evidence	Six-figure grid reference
Swanage tourist information office	*i*	031790

6. Look carefully at map A.
   a) Who has produced this map?
   b) What is the map trying to achieve?
   c) Identify five similarities and differences between map A and OS map-flap D, and record your findings in a two-column table.

7. Plan a five-day family holiday in Purbeck using map A, map-flap D and the tourist website: www.visit-dorset.com

8. Identify five ways that your holiday plan will support tourism jobs in the area.

How can I use charts and satellites to predict the weather?

Learning objectives

▶ To interpret data from weather charts and satellite images.

▶ To know and use the synoptic code.

Most countries in the world have developed an organisation responsible for weather forecasting. In the UK, it is called the Meteorological Office. Weather is measured by these organisations, and the results summarised in weather charts using the synoptic code – see A. Information for a weather station is summarised on a chart, like B, using a circle to represent the station. Observations at each individual station are placed around the circle, at a certain point, using the synoptic code.

Increasingly, satellites are also used to track the path of weather systems such as depressions and anticyclones. Satellite image C shows an anticyclone which dominated the weather over the UK on 29 June 2018. Anticyclones are high pressure systems. In summer, they produce hot, sunny, dry, calm days, and in winter, cold days, often with frost. Fog is common, particularly in coastal areas, and in valley floors.

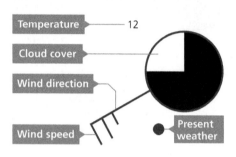

Temperature — 12
Cloud cover
Wind direction
Wind speed
Present weather

A A synoptic code

Symbol	Precipitation	Symbol	Cloud cover	Symbol	Wind speed
🌢	Drizzle	◯	Clear sky	◎	Calm
▽	Shower	◓	One oktas	◯—	1-2 knots
●	Rain	◕	Two oktas	◯⌐	5 knots
✱	Snow	◓	Three oktas	◯⌐	10 knots
△	Hail	◑	Four oktas	◯—	15 knots
⬙	Thunderstorm	◕	Five oktas	◯⌐	20 knots
⁖	Heavy rain	◕	Six oktas	◯▼	50 knots or more
✳	Sleet	◖	Seven oktas		
✱	Snow shower	●	Eight oktas		
=	Mist	⊗	Sky obscured		
≡	Fog	The sky is divided into eighths or oktas to record how much cloud cover there is.			

B Weather chart showing an anticyclone

Activities

1. Write definitions for the Meteorological Office and the synoptic code.

2. Write a definition of an anticyclone and describe the differences in the weather they bring in summer and in winter.

3. Look carefully at satellite image C.
   a) Write a sentence to explain why it is possible to see most of the UK in this image.
   b) Write a sentence to explain why east coast areas cannot be seen.

c) State two ways that satellite images help us to forecast the weather.

4. Compare satellite image C with weather chart B.
   a) What type of weather is being experienced along the east coast of mainland Britain? Give evidence from the chart and the satellite image.
   b) How is the temperature different along the east coast compared to the middle of the country?
   c) Account for this difference.

C Satellite image, 29 June 2018, NERC Satellite Receiving Station

29 June 2018

The weather forecast for the day stated:

... a slow-moving anticyclone will dominate the weather for the next few days, bringing fine and settled weather. Most places will be dry and sunny. Northern Scotland and the east coast will experience patches of fog.

43

How can I use satellite images to track depressions?

Learning objectives

▸ To interpret data from weather charts and satellite images of a depression.

▸ To understand how satellites can help us to predict the weather.

Weather satellite images are particularly useful for tracking and predicting the path of depressions. This is a type of weather system that often crosses over the UK. See weather chart A and satellite image B. It usually approaches from the west and generally brings unsettled weather with rain, dark clouds and sometimes strong winds.

A depression has three elements: warm front, warm sector and cold front. It forms as a result of warm air mixing and rising above surrounding cold air. A front is the boundary between a mass of warm and cold air. As the warm air is forced to rise over the cold air, condensation takes place, leading to cloud development, and eventually rain.

The pattern of clouds in a depression is clearly shown on satellite images as a mass of swirling clouds. Comparing successive satellite images helps track the movement of the weather system and predict likely movements to form future weather forecasts. High clouds are cooler than others and appear white on a satellite image. Low clouds, which are not so cool, appear grey. Speckled patterns of cloud may lead to rain showers. Eventually the cold front catches up with the warm front lifting the warm sector above the surface of the Earth. This is an occluded front.

A Weather chart showing a low pressure system

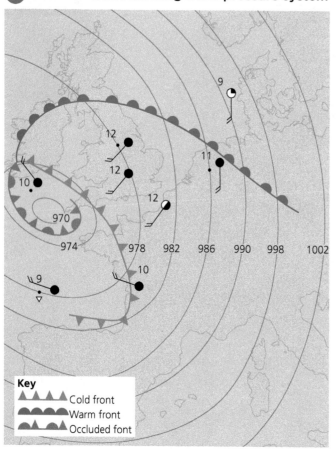

Key

⏶⏶⏶ Cold front

◠◠◠ Warm front

⏶◠⏶ Occluded font

Activities

1 What is a depression?

2 Look carefully at weather chart A.
   a) Draw a sketch of the depression and its position over the UK and Europe.
   b) Label the four components of the depression: warm front, cold front, warm sector and occluded front.
   c) Annotate the sketch to describe the weather over Britain using the synoptic code (shown in Lesson 4.1).

3 Look carefully at the satellite image of the depression, B.
   a) When was the image taken?
   b) Match the following with the numbers labelled on the image: high clouds; low clouds; speckles of clouds leading to showers of rain; centre of depression; warm front; cold front; warm sector; snow-covered mountains of the Alps; Spain; UK; Mediterranean Sea.
   c) How can satellite images help us to predict the weather?

4 Go to NASA's ZoomEarth website: https://zoom.earth. Look at today's satellite images of the world either as a globe or as a map. Identify the type of weather system over the UK and Europe.

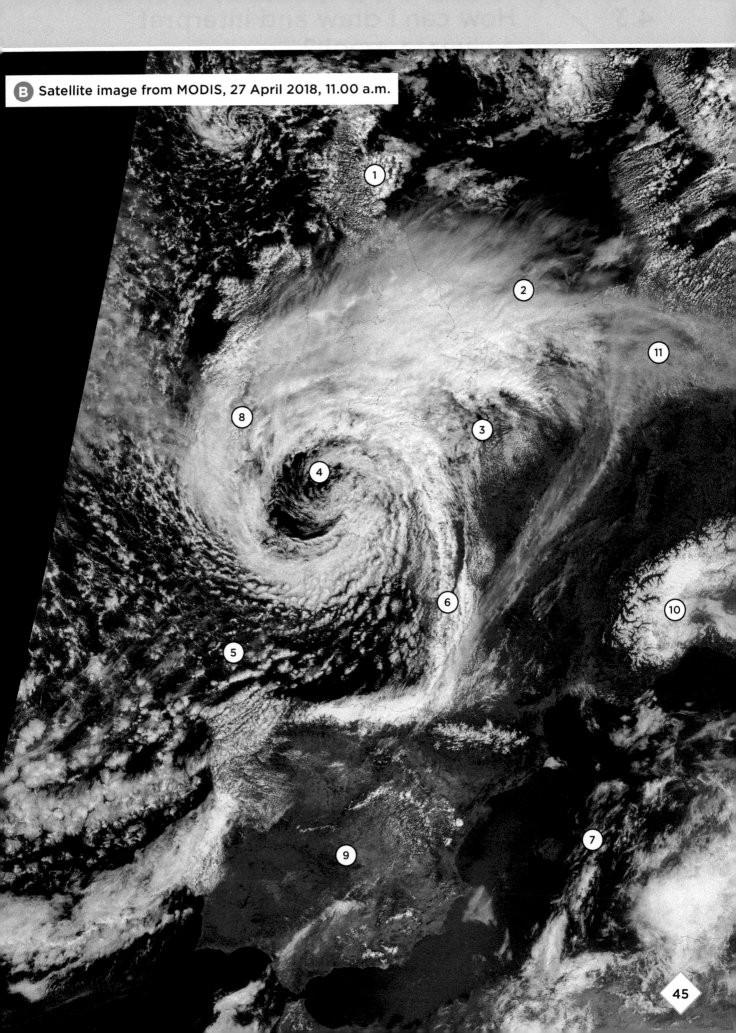

How can I draw and interpret a climate graph?

Learning objectives

▶ To be able to draw a climate graph.

▶ To be able to interpret a climate graph.

Organisations around the world collect climate data. This makes it possible to compare the climate of different places. Many locations have records going back more than 100 years.

When investigating the climate of different places, geographers put two graphs together to create a climate graph to make it easier to identify patterns between:

- **Temperature** shown on a line graph with monthly temperatures given in degrees Celsius (°C). The line on the graph is red.
- **Rainfall** shown on a blue bar chart with average monthly rainfall given in millimetres. The bars are shaded blue.

A How to draw a climate graph

Step 1: Look at the data. Use the wettest month and the months with the highest and lowest temperatures to help you choose a suitable scale for the axis. In some cases, the temperature axis may need minus readings below the 0° gradation.

Step 2: Draw three axes:
- temperature on the left vertical
- rainfall on the right vertical
- months on the horizontal.

Label each axis.

Step 3: Plot the rainfall figures as a bar chart. Check for accuracy and then shade each bar in blue.

Step 4: Plot the temperature data, making sure each cross or dot is placed in the centre of the month. Use a red pencil to join the points with a smooth, red curve.

Step 5: Add a heading that includes the name of the place being graphed and its latitude, longitude and elevation – as these factors affect the climate of a place.

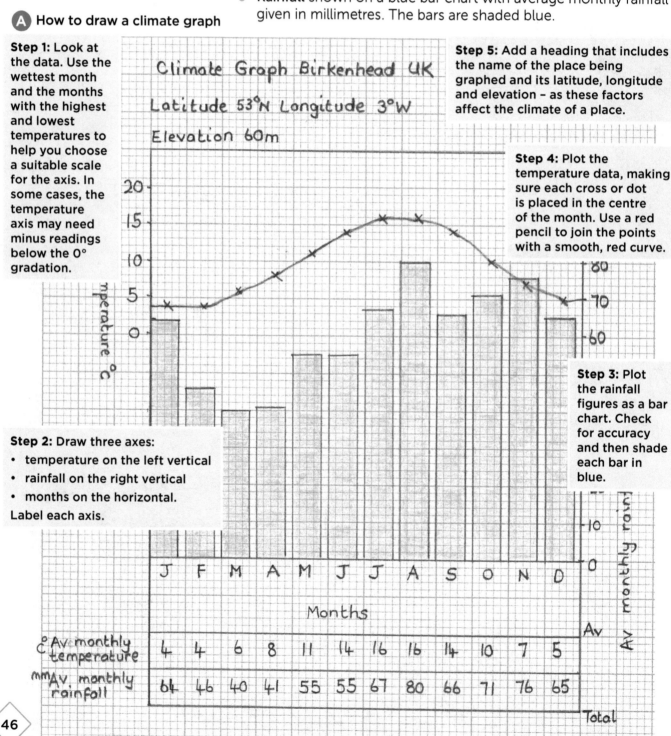

Climate Graph Birkenhead UK
Latitude 53°N Longitude 3°W
Elevation 60m

	J	F	M	A	M	J	J	A	S	O	N	D	
°C Av. monthly temperature	4	4	6	8	11	14	16	16	14	10	7	5	Av
mm Av. monthly rainfall	64	46	40	41	55	55	67	80	66	71	76	65	Total

Average monthly temperature	Description
Above 30°C	Very hot
21–30°C	Hot
10–20°C	Warm
0–9°C	Cool
–10–0°C	Cold
Below –10°C	Very cold

Annual temperature range	Description
Above 30°C	Very large
15–30°C	Large
5–15°C	Moderate
Below 5°C	Small

Average monthly rainfall	Description
Above 150 mm	Very wet month
50–150 mm	Wet month
Below 50 mm	Dry month

Interpreting a climate graph to describe the climate of a place

Once you have drawn a climate graph, you can use it in various ways to describe and interpret the climate.

- Begin by calculating the average monthly temperature, range of temperature and annual rainfall, and quote these figures.
- Look at the overall shape of the climate graph. Consider whether the temperature line graph is steep or gentle, or if it changes at different times of the year. See if there is a similar or different pattern of rainfall.
- Look for extremes – quote the highest and lowest months of the year for temperature and rainfall.
- Use the descriptive categories shown in tables B, as you describe the climate.
- See if you can identify seasons when most and least rain falls or when highest and lowest temperatures are experienced.

Calculating key data

Average monthly temperature = add together the temperature for each month and divide the total by 12 (the number of months in a year).

Temperature range = the difference between the lowest monthly temperature and the highest.

Annual total rainfall = add together the rainfall for each month.

Activities

1 What is a climate graph?
2 Look carefully at climate graph A for Birkenhead.
   a) What do the bar and line graphs show?
   b) Calculate the average monthly temperature, the temperature range and the total annual rainfall.
   c) Write a paragraph to describe the climate of Birkenhead, using the descriptive words in tables B and the guidance on interpreting a climate graph.
3 Tables C and D show climate data for Tromsø and Santiago.
   a) Draw a climate graph for each place.
   b) Then repeat Questions 2b) and 2c) for Tromsø and Santiago.

C Climate data for Tromsø, Norway

Tromsø, Norway, 70°N 19°E, 100 m												
	J	F	M	A	M	J	J	A	S	O	N	D
Av. temp, °C	–4	–4	–3	0	4	9	13	11	7	3	0	–2
Rainfall, mm	96	79	91	65	61	59	56	80	109	115	88	95

D Climate data for Santiago, Chile

Santiago, Chile, 33°S 71°W, 520 m												
	J	F	M	A	M	J	J	A	S	O	N	D
Av. temp, °C	21	20	18	15	12	9	9	10	12	15	17	20
Rainfall, mm	3	3	5	13	64	84	76	56	31	15	8	5

4 Describe the absolute and relative location of Birkenhead, Tromsø and Santiago, using an atlas, and mark their location on an outline map of the world.
   a) Write a paragraph comparing the climate of the three locations. Identify how they are similar and how they are different.
   b) Consider where each are located in the world, and their height, and try to explain the differences in climate.

Learning objectives

▶ To use an atlas map of Russia.

▶ To locate places using latitude and longitude.

▶ To measure distance between places on an atlas map.

▶ To be able to use GNSS to investigate a place.

When geographers investigate a place or country, they first consider its location. This involves a range of skills, including using atlas maps. Today we can also use modern technology – computers and satellites to do this, as part of a Global Navigation Satellite System (GNSS). In this lesson you will use these to investigate the location of Russia.

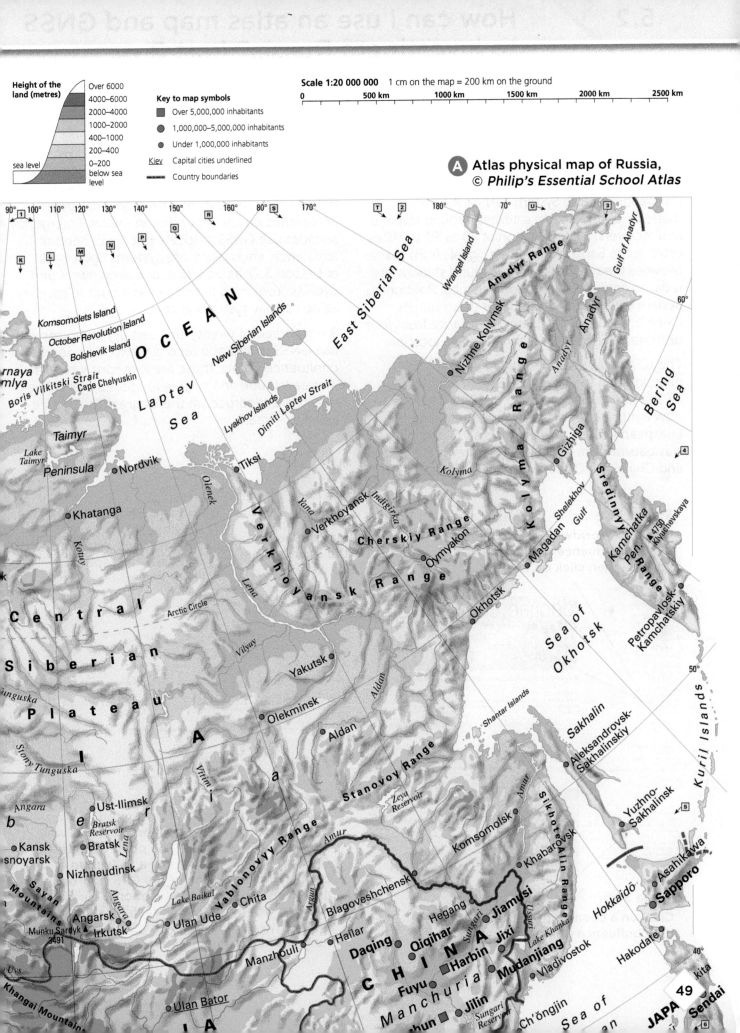

5.2

How can I use an atlas map and GNSS to investigate Russia? Part 2

Global Navigation Satellite Systems

People have looked to the skies to find their way since ancient times. Sailors used the constellations of stars to navigate. Today, to help us navigate, instead of stars we use satellites, as part of a Global Navigation Satellite System (GNSS). A GNSS is made up of three parts: satellites, ground stations and receivers. A constellation of 20 to 30 satellites orbiting the Earth, transmitting signals from space, provides their exact position. Ground stations use radar to confirm the location. A Global Positioning System (GPS) receiver listens for these signals. Once the receiver calculates its distance from four or more GPS satellites, it can calculate location.

The USA's NAVSTAR GPS launched the first experimental satellite in 1978, with a full 24 GPS satellites orbiting Earth by 1994. There are now a number of other systems, including the European Union's Galileo, Russia's Global'naya Navigatsionnaya Sputnikovaya Sistema (GLONASS) and China's BeiDou Navigation Satellite System.

Who uses GNSS?

The first system developed in the USA was initially for the military; in fact, it is still operated by the US Air Force. Some of the other systems used today are for civilians – they can be used to map forests, help farmers harvest their fields and navigate aeroplanes on the ground or in the air. Emergency services use GNSS to locate people in need of assistance; ships use these systems (look back at Lesson 3.2), as do trains, and even hikers and cyclists. Your mobile phone has a built-in receiver, as do sat-nav systems in cars.

A new breed of twenty-first century explorers use GNSS to visit and record the world's degree confluence points; these are locations where a degree latitude and degree longitude meet. The results are recorded on the Degree Confluence website – see B.

Stage 1

Click on 'Worldwide maps' to see all the confluence points, which you can then click on to visit.

Stage 2

Click on 'Search' and you can enter the coordinates of a place to visit.

B **How to use the Degree Confluence Project website to locate places. Website homepage:**
http://confluence.org/

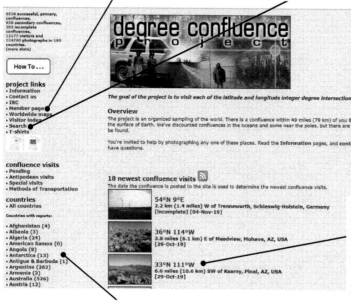

Stage 3

Click on a country to see a map of all the confluence points for that country.

Stage 4

Each confluence point is presented with a photo and its latitude and longitude, as well as where it is and when the point was visited. Click on the location to see a collection of photos, including one of the GPS as proof, and a description of the location. There is also a link to Google Maps for each location. An interactive compass allows you to move to the next confluence point in any direction.

1 Look carefully at atlas map A on pages 48–49.
   a) Give the latitude and longitude of the following cities in Russia: Moscow, Irkutsk, Omsk, St. Petersburg, Murmansk, Vladivostok.
   b) Use the key to help you name the two largest cities in Russia.
   c) Name the ocean that borders the North of Russia.
   d) Name the 14 countries that border Russia.
   e) Name the mountain range that runs from north to south, dividing Russia into two parts.
   f) Use the linear scale to find out how wide Russia is in kilometres from west to east.

2 Make a copy of the matrix below and complete it by measuring the straight line distance between the Russian cities shown on atlas map A.

St. Petersburg
Omsk
Moscow
Yakutsk
Vladivostok
Perm

3 What is a Global Navigation Satellite System?

4 a) Use an online search engine such as Google to search for the following video and watch it: 'How does a global navigation satellite system work?'
   b) Write a paragraph to explain how a GNSS works.

5 Look back at Lesson 3.2, Question 6. How do ships use GNSS?

6 Identify five other ways people use GNSS.

7 Go to Google Maps, either on your mobile phone or on the internet.
   a) Enter your postcode.
   b) Get directions and find out how far each Russian city, shown in the matrix for Question 2, is from your home.
   c) Record the results on another copy of the matrix in Question 2.

8 Visit the Degree Confluence website and use B to help you navigate the website.
   a) Go to the map of Russia.
   b) Look at atlas map A and select three different places across Russia.
   c) Click on the website map of Russia, and visit these three locations, look at the photos, and read the descriptions.
   d) At each location, download a photo, and annotate it using the five enquiry questions (see page 2).

9 Reflect on this lesson, and write five sentences to summarise what you have discovered about Russia.

10 Look at the vision statement, flap C. Which aspects of becoming a geographer have you made progress with in this lesson?

How can I use a computer to draw graphs?

Learning objectives

▶ To be able to use a computer spreadsheet to draw graphs.

▶ To investigate statistical data about Russia.

Graphs can be drawn using computer spreadsheet software. These are very easy and quick to produce. It is also possible to select different types and design styles of graphs, or charts as they are called in the program. It is also possible to copy and paste a completed chart into a Word file, where the user can add their own interpretation of the chart. In this lesson, using a computer, you can follow the steps shown to create charts for different data about Russia.

A How to use a computer to draw a graph

	A	B	C	D
	Primary	Secondary	Tertiary	Quaternary
	1%	18%	57%	24%

Step 1: Enter the data into the spreadsheet. Type the headings for each category of data in the first row and the value in the second row.

Step 2: Highlight the cells, select 'Insert' from the menu bar, and finally 'Sheet'. This will create a separate sheet in the spreadsheet workbook for your chart.

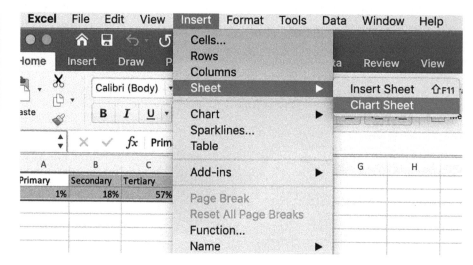

Step 3: Click on 'Graph' and select the type of chart you want to use. Economic sectors are shown as a pie chart. At this stage, you can also select a presentation style for your chart.

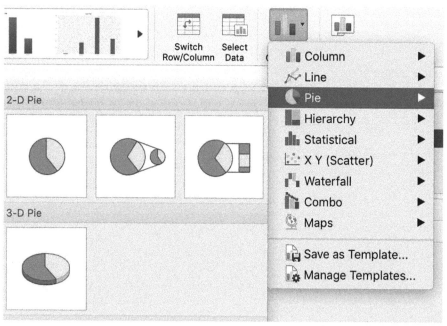

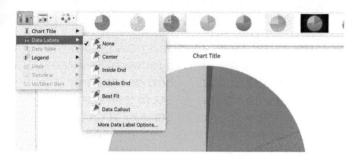

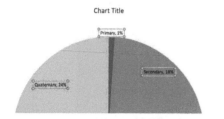

Step 4: You can start to label and edit the chart, by clicking on 'Add Chart Element'. Here you can add data labels, and include or delete a legend. Pie charts are best presented without a legend, or key. They are best presented with labels for the data category and value for each segment of the pie chart.

Step 5: Choose a suitable font type and size for the labels, and the chart title. You can also click on the labels and title and move them to a position you are happy with.

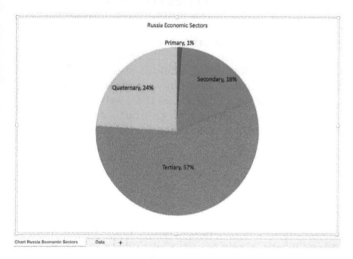

Step 6: Rename the separate sheets in your spreadsheet workbook, in the bottom left-hand corner of the sheet. This will make it easier to find charts and data as you build up a number of charts as you investigate Russia.

B Russian exports

Exports from Russia	%
Crude oil and gas	36
Refined oil products	18
Food	6
Chemicals	7
Wood and paper	3
Machinery and equipment	13
Metals	11
Other goods	6

Activities

1 Follow the steps and use the data in A, Step 1, to create a chart to show the percentage of the workforce employed in each economic sector in Russia.
2 Copy and paste the chart into a Word document.
3 Write a paragraph in the Word document to describe the economic structure of Russia.
4 Compare the economic structure of Russia with that of the UK, shown in Lesson 3.1, chart A.
5 Enter the data shown in table B into a spreadsheet, and follow the steps above to create a chart for the data.
6 Copy and paste your chart into a Word document, and underneath it describe what it shows.

Learning objectives

▶ To be able to draw a cross-section from an OS map.

▶ To know the key features of a drainage basin.

Turn back to and read Lesson 1.8, which explains how height is shown on OS maps. Contour lines are used to represent a three-dimensional shape of the Earth's surface. A good geographer develops the ability to visualise contour patterns in three dimensions and drawing cross-sections helps develop this ability. A cross-section is a view of a landscape as it would appear if sliced open, a side view of a landscape. OS map A shows a 1:25 000 OS map extract near the source of the River Derwent. Styhead Gill is one of two tributaries that form the River Derwent. The contour lines are very close together, and follow the course of the Gill. This is the typical contour pattern of a river valley.

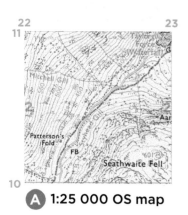

A 1:25 000 OS map

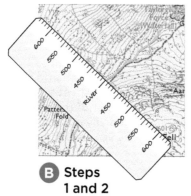

B Steps 1 and 2

© Crown copyright and database rights 2020 Hodder Education under licence to Ordnance Survey

Step 1: Place the straight edge of a strip of paper on the OS map between the two locations either side of the valley.

Step 2: Mark off every point where a contour line crosses the edge of your paper. Record the height of each contour. (Remember, contours are drawn every 10 metres. Some have their height marked on them. Every fifth contour – every 50 metres – is shown with a thicker line.)

How to draw a cross-section

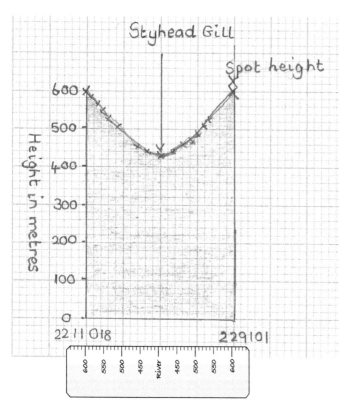

C Steps 3 to 7

Step 3: Draw a grid for your cross-section. Use a vertical scale of 1 cm for every 100 metres – the length of the horizontal axis is formed by the length of the cross-section.

Step 4: Place your piece of paper on the bottom edge of your grid. Draw a dot on the correct height line on your grid, directly above each contour line marked on your piece of paper.

Step 5: Join the dots together with a smooth, freehand curved line. Shade in below the line.

Step 6: Label the main features along the line of the cross-section.

Step 7: Add a title to the vertical axis, and write the six-figure grid references for the two end points.

A drainage basin

A drainage basin is a bowl-shaped depression which feeds a river. Inside the basin is a network of streams and rivers where water flows from the start or source of a river to its mouth, where it ends. A small stream that feeds water into the main river, like Styhead Gill, is called a tributary. Where they meet is called the confluence. The dividing line between neighbouring drainage basins is called the watershed.

How to make a pop-up drainage basin

A pop-up drainage basin will help you to better understand how a drainage basin works by seeing it in three dimensions.

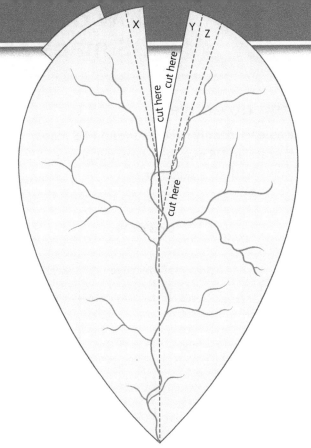

D Template for a pop-up drainage basin

Step 1: Make a large copy of diagram D.

Step 2: Label the following on your diagram: source, main river, tributary, mouth, watershed, confluence.

Step 3: Glue X over flap Y.

Step 4: Fold inwards along the dotted lines.

Step 5: Fold Z over and glue onto flap X.

Step 6: Fold the model along the dotted line which is not glued.

Step 7: Glue this section into the centre of a double page in your exercise book. Make sure that your model folds flat when your book is closed. When you open your book, the model should pop up!

Activities

1 What is a cross-section?
2 How does a cross-section help you to visualise contour patterns in 3D?
3 Look carefully at map A.
  a) Draw a sketch map to show the contour pattern along the river valley for Styhead Gill.
  b) Write three sentences to describe what this river valley is like.
  c) Describe how cross-section C shows a different view of the valley.
4 Look at map-flap E. The River Derwent flows from the north-east through the town of Cockermouth.

a) Compare the river valley with the tributary of the Derwent shown on map A. Write three sentences to describe how the two are different.
b) Draw a cross-section of the Derwent valley from the A595 road at 134340 to the footpath at 152323. Use the guidance to help you.
c) Compare your cross-section with C. Write three sentences to describe how the two valleys are different.
5 Make a copy of template D and follow the instructions to create your own pop-up drainage basin.
6 Write a paragraph to describe the key features of a drainage basin.

Learning objectives

▶ To be able to identify river features on OS maps.
▶ To compare an OS map with an aerial photo.
▶ To be able to draw a sketch map.

Although no two rivers are exactly the same, many share a similar pattern in their long profile, which shows changes in gradient from source to mouth, and cross profile, the slopes across the valley sides like cross-sections. The gradient of the long profile is steep in the upper course of the river and becomes increasingly less steep further downstream. As a result, a river has different characteristic features along its course, shown in diagram A.

Meanders

Meanders develop in a river in the middle and lower courses as the river slows down. They are bends in the course of the river. On the outside of the bend, the river flows faster eroding the bank – this is called a river cliff; on the inside of a bend, the river moves more slowly, depositing sand and pebbles – this is called a slip-off slope. Meanders (see B) slowly move across valleys forming wide and flat valley floors, called a flood plain. It is possible to identify the features of a meander by comparing OS maps and aerial photo D.

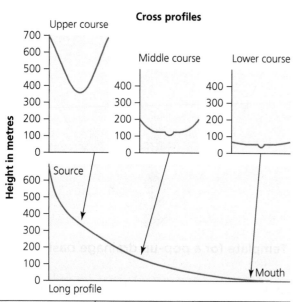

Upper course	Middle course	Lower course
Steep sided, narrow and deep V-shaped valley. River channel takes up most of the valley floor.	Valley broadens out. It now has a small flood plain. Valley sides are not as steep.	Large expanse of flat flood plain on either side of the river.

A How a river valley profile changes downstream

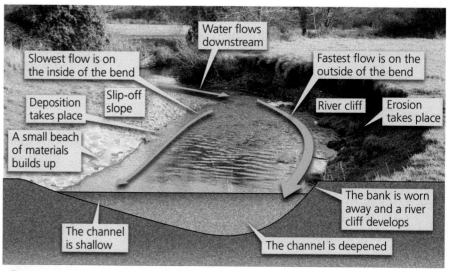

B Features of a meander

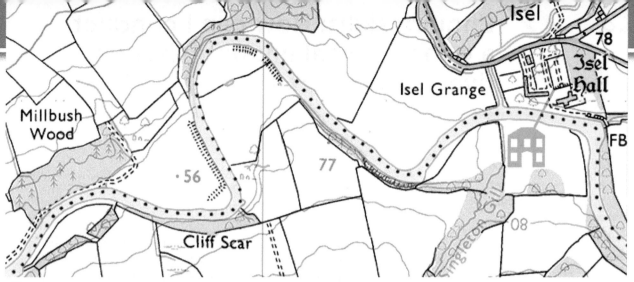

C 1:25 000 OS map of part of River Derwent valley

© Crown copyright and database rights 2020 Hodder Education under licence to Ordnance Survey

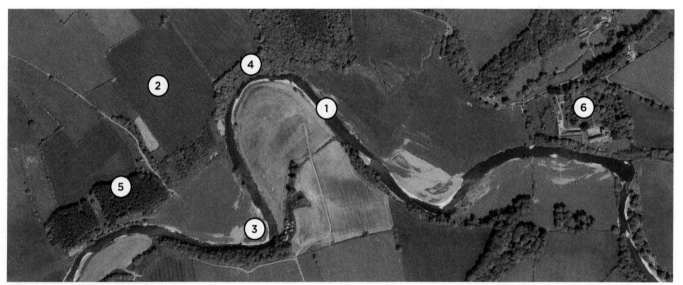

D Aerial photo of River Derwent valley from Getmapping

Activities

1 Look carefully at diagram A. Write a paragraph to describe the typical three stages of a river, and how the width and depth of the valley changes from source to mouth.

2 Go back to your pop-up drainage basin from Lesson 6.1. Annotate where you think each stage of the river is on your basin.

3 Compare OS map C and aerial photo D, map-flap E and diagram B.
   a) Locate map C and photo D on map-flap E, giving the four grid squares they are in.
   b) Identify the river features labelled 1–4, and use map-flap E to give their six-figure grid references.

c) Identify land uses 5 and 6 and give their six-figure grid references.

4 Look carefully at diagram B.
   a) Draw a sketch cross-section of the meander from the river cliff to the slip-off slope.
   b) Annotate your sketch to show these two banks and why erosion and deposition is happening.

5 Compare your sketch cross-section with C, D and map-flap E.
   a) Label the six-figure grid reference for a location on map C where you think the feature and processes in your sketch cross-section are happening.
   b) Write three sentences to justify your choice, using and including evidence.

Enquiry skills: How can I conduct a geographical enquiry about a river flood? Part 1

Learning objectives

▶ To identify enquiry questions.

▶ To be able to interpret a range of geographical data as part of an enquiry about a flood.

▶ To be able to reach conclusions consistent with the evidence and enquiry questions.

Cockermouth is a town in the county of Cumbria in the Lake District. It is located at the confluence of two rivers. The town has always been prone to flooding, with 15 flood events recorded since detailed records began in 1761. On 5 and 6 December 2015, the region experienced a major flooding event, affecting numerous towns in the county, the result of a major storm. In Cockermouth, 594 properties were flooded. As a result of a severe flood in 2009, flood defences for the town were improved by the Environment Agency. Unfortunately, these were over-topped in 2015, when the water level exceeded the height of the defences. In this double lesson, you will conduct an enquiry about the 2015 flood event.

A Satellite image of the 2015 Storm Desmond from EUMETSAT

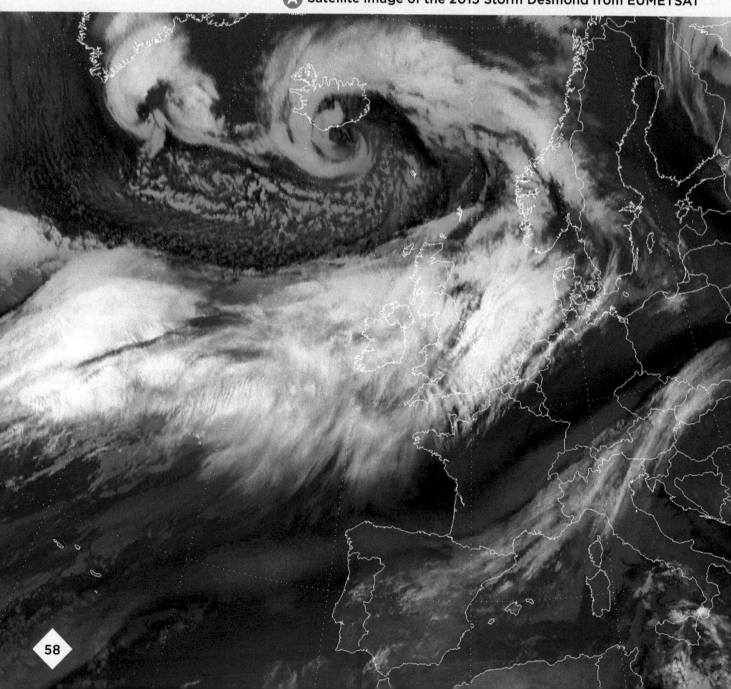

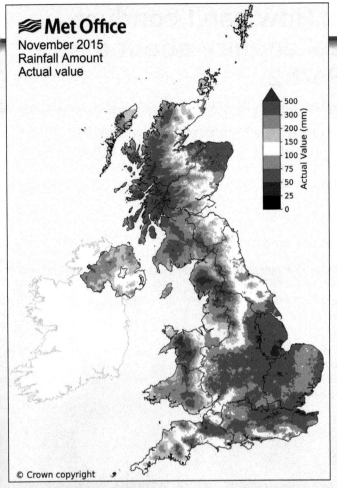

B Met Office rainfall map for November 2015

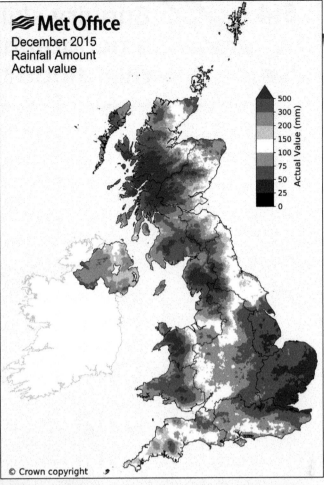

C Met Office rainfall map for December 2015

Storm Desmond

The severe flooding in Cumbria was principally the result of Storm Desmond. This was a deep Atlantic low-pressure system. Desmond had strengthened and deepened as it travelled eastwards across the Atlantic, gaining a lot of moisture from the Caribbean as it passed over. It became a system of very moist air several thousand kilometres long and a few hundred kilometres wide (see satellite image A). Storm Desmond dragged this moisture-rich air over Cumbria. As it rose over the mountains, the air cooled, leading to condensation of the water vapour. The cold front became stationary over Cumbria. The cold air appears blue on image A, and surrounding warm air green. This all led to very intense, prolonged and record-breaking rainfall – see C and D. Unfortunately, the ground in the north of England was already very wet. Many parts of north-west Britain had already recorded more than twice the monthly average rainfall during November (see B), the result of a succession of earlier, similar but less intense, storms like Desmond.

D Key weather statistics about Storm Desmond, Met Office

Highest 48-hour rainfall record for UK (405.0 mm at Thirlmere, Cumbria, in just 38 hours)

Wettest calendar month on record for UK (191 per cent of December average)

Highest 24-hour rainfall record for UK on 5 December (341.4 mm at Honister Pass, Cumbria)

The rain isn't letting up, and flood waters are rising at an alarming rate. Four of the six bridges in Cockermouth weren't passable at 4 pm, and it's getting worse now. People's homes and businesses are at risk. Dreadful.

We stayed in our house because after the 2009 flood we bought flood defences for our home so we thought we'd better stay in and try and mop up any leaks that come in, but we just got overwhelmed and we ended up leaving our house.

E Eyewitness accounts of the Cockermouth flood

6.4

Enquiry skills: How can I conduct a geographical enquiry about a river flood? Part 2

> The flood defences were designed for a one in 100 year event and it's six years since we had the last one. We were sort of surprised that we got one so soon.

F Local resident's view

The Environment Agency is a government body, whose responsibilities include managing the risk of flooding in the UK, as well as making people aware of flood risk and advising them on how to protect themselves. In 2018, they published new online guidance on how to cope with flooding: https://flood-warning-information.service.gov.uk/what-to-do-in-a-flood/getting-a-flood-alert

GIS map H is part of this guidance.

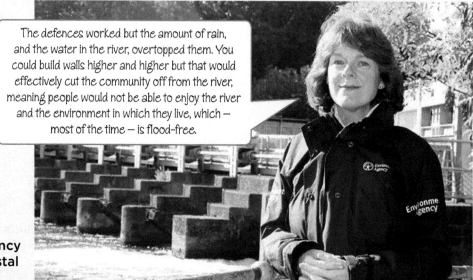

> The defences worked but the amount of rain, and the water in the river, overtopped them. You could build walls higher and higher but that would effectively cut the community off from the river, meaning people would not be able to enjoy the river and the environment in which they live, which — most of the time — is flood-free.

G Former Environment Agency director of flood and coastal risk, Alison Baptiste

H Environment Agency online flood risk map.
© Crown copyright and database rights 2020 Hodder Education under licence to Ordnance Survey

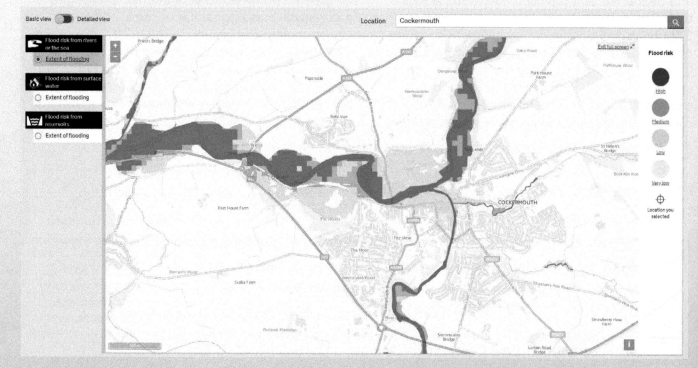

Activities

1 Read through the geographical data about the Cockermouth flood.

   a) Which of the '4 Ws and an H' enquiry questions (introduced in Lesson 1.1) do you think you could use and answer about this flood?

   b) Think up several questions of your own to investigate the flood.

2 Look carefully at map-flap E.

   a) Name the two rivers which flow through Cockermouth.

   b) Give the six-figure grid reference for the confluence of these two rivers.

   c) Use evidence provided on the map to write three sentences to explain why the town is prone to flooding.

3 Investigate the ArcGIS storymap about the Cockermouth flood of 2015.

   a) Go to the ArcGIS website homepage: https://schools.esriuk.com/teaching-resources/#-
   Enter 'Cockermouth Flood 2015 case study', and click on 'Cockermouth December 2015 flood case study package'.

   b) Click on each of the circle pages of the story, and record evidence about the nature of the flood. Some pages include videos, maps, graphs, GIS maps and activities for you to complete.

   c) Write a paragraph summarising what you have discovered about the flood.

4 Look carefully at satellite image A of Storm Desmond.

   a) What type of weather system was Storm Desmond, and when did it cross over the UK?

   b) On an outline map of the UK and Europe, draw a sketch map of Storm Desmond.

   i) Label the centre of the storm, and shade cold air blue and the surrounding warmer air green.

   ii) Draw an arrow on your sketch to show the direction in which the weather system was moving.

   iii) Annotate your sketch to explain why the storm led to record-breaking rainfall over Cumbria.

   iv) Annotate the evidence in D on your sketch.

5 Look carefully at maps B and C.

   a) What are these two maps showing?

   b) Compare each map with map B in Lesson 2.1, and write a paragraph to describe the distribution patterns they show. (Look back to Lesson 2.1 for guidance on how to describe patterns.)

   c) Write a sentence to explain how the high levels of rainfall in Cumbria in November contributed to the floods in December 2015.

6 Read the eyewitness accounts of the flood E and views you discovered on the ArcGis storymap.

   a) Write a paragraph to describe what the flooding was like for the people of Cockermouth.

   b) Read F, the viewpoint of a Cockermouth local resident. Write a sentence to explain why it is a problem to take the term 'a one in a 100 year event' literally.

7 Read view G and compare it with D and E. Write a paragraph to explain why it is difficult for the Environment Agency to develop flood defences that will always protect the people in towns like Cockermouth.

8 Compare GIS map H with map-flap E.

   a) Give the four-figure grid references of the areas of Cockermouth that have the highest flood risk.

   b) Write three sentences to describe the locations of the areas of Cockermouth that have the greatest risk of flooding.

9 Conclusion: Now go back to your enquiry questions from Question 1. Look through all you have discovered about the 2015 Cockermouth flood. Write a paragraph that answers your enquiry questions, summarising what you have learnt.

10 Look at the vision statement, flap C. Which aspects of the vision do you think you have made progress with in this enquiry?

How can I use choropleth maps and cartograms to investigate development patterns?

Learning objectives

▶ To use a choropleth map to investigate patterns of development.

▶ To compare a choropleth map with a cartogram.

Development is an important concept in geography. It basically means people reaching an acceptable standard of living or quality of life. Each country in the world is at a different stage of development. Geographers investigate and compare distribution patterns of development by plotting data on maps. Map A shows a choropleth map, plotting Gross National Income (GNI) per capita – an indicator of development. A choropleth map is a thematic map in which areas are shaded according to a range of data that is linked. Different colours are used to show categories of the data. Each area is shaded using one colour, where the darker shades represent high numbers and the lighter shades represent low numbers.

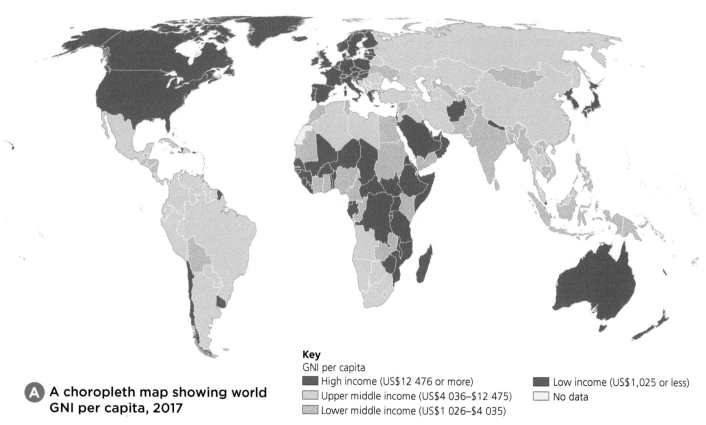

Key
GNI per capita
⬛ High income (US$12 476 or more)
⬜ Upper middle income (US$4 036–$12 475)
◻ Lower middle income (US$1 026–$4 035)
⬛ Low income (US$1,025 or less)
⬜ No data

A A choropleth map showing world GNI per capita, 2017

Activities

1 Write three sentences to explain choropleth maps.

2 Create a two-column table and summarise the advantages and disadvantages of choropleth maps.

3 Look carefully at map A.
  a) What statistic is plotted on the map?

  b) Look back at Lesson 2.1 to remind yourself how to describe a distribution pattern. Then write a paragraph describing the world distribution pattern you can see on this map.

  c) Divide the world into four zones according to the dominant four GNI per capita gradings given in the key.

Advantages and disadvantages of choropleth maps

Choropleth maps make it easier to see geographical patterns rather than looking through a table of statistics. However, choropleth maps have a number of disadvantages: they give a false impression that there is a sudden change at area boundaries; variations in the data at different scales are hidden – for example, in map A there is a suggestion that the GNI per capita is the same for the whole country; and the map gives no indication of the different sizes of population in each country.

Cartograms

A cartogram combines statistical information with geographic location. A cartogram uses statistics to change a place's area, resizing it in proportion to the statistic, rather than the area of the place. Map B, for example, shows countries resized according to their Gross Domestic Product (GDP), another measure of development. The colours show regions of the world. The cartogram has been produced by Worldmapper – you can investigate many more on their website: https://worldmapper.org

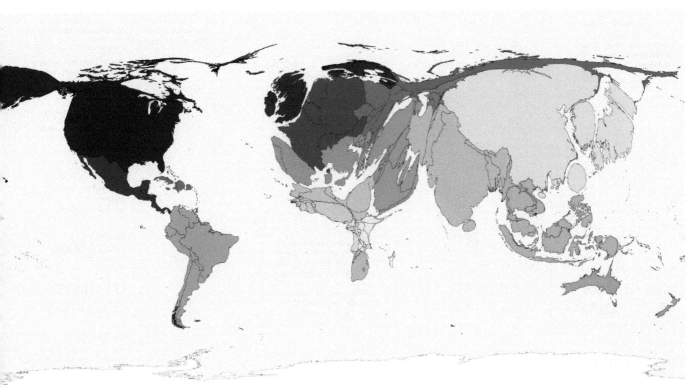

B A cartogram showing GDP, 2018, Worldmapper

4 Write three sentences to explain what cartograms are.

5 Look carefully at map B.
   a) Look at the size and shape of the British Isles. Explain how and why its size and shape have changed from map A.
   b) Which continents have changed shape the most? Account for this change.
   c) Repeat Question 3, but using map B.

6 Compare maps A and B.
   a) How is the data presented in map B different from the data presented in map A?
   b) Write two sentences to explain which map you find the most useful to identify patterns of development.
   c) Identify an advantage and disadvantage of the cartogram map, B.

How can scatter graphs and Gapminder help me to understand development?

Learning objectives

▶ To use a Gapminder bubble chart to investigate development.
▶ To identify relationships between data sets.

Scatter graphs

Another tool that geographers can use to compare development data between countries is a scatter graph. A scatter graph can be used to show relationships between any two sets of data. To plot a scatter graph, you first need to select two data sets, one shown on the x-axis, and one shown on the y-axis. A dot or cross is plotted at the intersection of the two sets of data for each country. A line of best fit helps to show correlations, or patterns across the data – see the examples, B–D. Using lines of best fit, geographers can draw conclusions about trends and patterns.

A Gapminder bubble chart tool guide

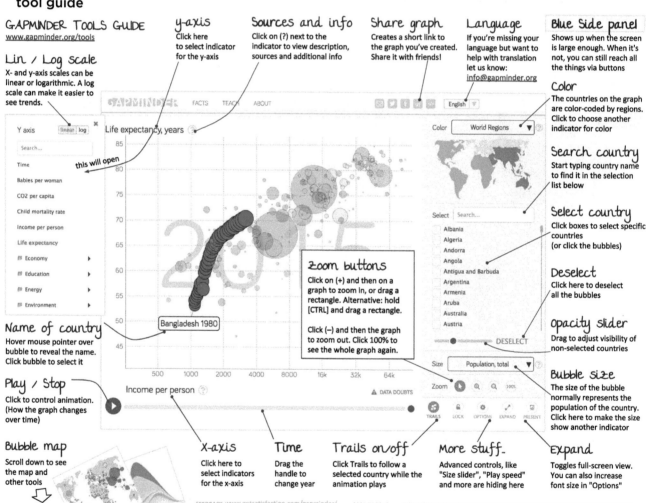

GAPMINDER TOOLS GUIDE
www.gapminder.org/tools

Lin / Log scale
X- and y-axis scales can be linear or logarithmic. A log scale can make it easier to see trends.

y-axis
Click here to select indicator for the y-axis

Sources and info
Click on (?) next to the indicator to view description, sources and additional info

Share graph
Creates a short link to the graph you've created. Share it with friends!

Language
If you're missing your language but want to help with translation let us know: info@gapminder.org

Blue Side panel
Shows up when the screen is large enough. When it's not, you can still reach all the things via buttons

Color
The countries on the graph are color-coded by regions. Click to choose another indicator for color

Search country
Start typing country name to find it in the selection list below

Select country
Click boxes to select specific countries (or click the bubbles)

Deselect
Click here to deselect all the bubbles

Zoom buttons
Click on (+) and then on a graph to zoom in, or drag a rectangle. Alternative: hold [CTRL] and drag a rectangle.
Click (–) and then the graph to zoom out. Click 100% to see the whole graph again.

Name of country
Hover mouse pointer over bubble to reveal the name. Click bubble to select it

Opacity slider
Drag to adjust visibility of non-selected countries

Bubble size
The size of the bubble normally represents the population of the country. Click here to make the size show another indicator

Play / Stop
Click to control animation. (How the graph changes over time)

Bubble map
Scroll down to see the map and other tools

X-axis
Click here to select indicators for the x-axis

Time
Drag the handle to change year

Trails on/off
Click Trails to follow a selected country while the animation plays

More stuff...
Advanced controls, like "Size slider", "Play speed" and more are hiding here

Expand
Toggles full-screen view. You can also increase font size in "Options"

FEEDBACK www.getsatisfaction.com/gapminder/   MARCH 2014   THIS GUIDE IS ADAPTED FROM AN ORIGINAL IDEA BY WWW.JUICYGEOGRAPHY.CO.UK

Gapminder bubble charts

Gapminder is an organisation that has designed data visualisation tools to present complex development data in a way that people can enjoy and better understand it. The bubble chart, shown in A, is like a scatter graph, but it visualises data in multiple ways:

● Each country in the world is shown as a bubble.
● The size of the bubbles represents the population size.
● The colour represents regions of the world.

On the chart shown in A, the x-axis shows GDP per capita in dollars per person per year, for each country. The y-axis shows life expectancy, from 25 to 85 years. When you click 'play', the bubbles are animated, showing how the data for each country changes through time.

Scatter graphs with different correlations

Positive correlation	Negative correlation	No correlation
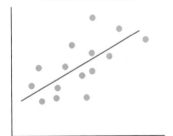	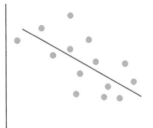	

B Line of best fit runs from bottom left to top right, so as one variable increases so does the other

C Line of best fit runs from top left to bottom right, so as one variable increases the other decreases

D Dots are distributed all over the graph, showing there is no relationship between the variables

Activities

1 What is the main use of a scatter graph?

2 Write two sentences to explain how a Gapminder bubble chart is like a scatter graph.

3 Look carefully at A, which shows how to use the Gapminder bubble chart.
   a) Identify the data shown on the x-axis and y-axis of the chart.
   b) Write a list to show what the following represent on the chart: a bubble, colour of bubble, size of bubble.
   c) Watch the video on the Gapminder website, where Hans Rosling explains what the graph shows: www.gapminder.org/answers/how-does-income-relate-to-life-expectancy/
   d) Go to the Gapminder website: www.gapminder.org
      Select 'Tools', and the graph shown in A will appear. Play the animation.
   e) Use the video and the bubble chart to write a paragraph to explain how the countries of the world have developed over the last 200 years. Identify things that surprise you about the data in the graph.

4 Compare the graph in A with B–D.
   a) What type of correlation exists between life expectancy and income per person?
   b) Write three sentences to explain this pattern.

5 Return to the Gapminder website and use A to help you.
   a) Click on the y-axis to change the data. Select 'Economy' and 'Sectors' and then 'Agriculture'. The bubble chart will change to show the data relating to agriculture and income.
   b) Compare the graph with B–D. What type of correlation exists between agriculture and income per person?
   c) Write two sentences to explain this pattern.

6 Use A to help you further explore the bubble charts on the website to better understand development.

How can I develop a fact-based world view?

Learning objectives

▶ To use the Gapminder Dollar Street website to investigate how people live in different stages of development.

▶ To use data to challenge a misconception called 'Gap Instinct'.

▶ To use ideas presented in a non-fiction book.

In 2018, the creators of Gapminder published a book, *Factfulness*. As you can see from the cover, A, it challenges our view of the world. The book identifies ten instincts that lead to people misinterpreting facts about the world. In this lesson, you will be introduced to the first instinct in the book, the 'Gap Instinct', which, the authors state, makes us divide things into two groups.

People tend to believe that the countries in the world can be divided into rich and poor, with a gap between the two groups. Hans Rosling, the founder of Gapminder, believes, 'There will always be the richest and the poorest ... The majority is usually to be found in the middle.' He argues it is more useful to divide the world into four income levels, expressed in terms of dollar income per day – see B. Each figure in the chart represents 1 billion people. Human history started with everyone at Level 1. Just 200 years ago, 85 per cent of the world's population was still at Level 1. Today the majority is now spread across Levels 2 and 3, with the same range of standards of living as people in the UK and USA had in the 1950s.

A *Factfulness* by Hans Rosling, book cover

Level 1 $2   Level 2 $8   Level 3 $32   Level 4
Income per person in dollars per day adjusted for price differences

B The four levels of income identified by Hans Rosling

C Dollar Street website tool guide, www.gapminder.org/dollar-street

Home page: shows families on Dollar Street, poorest to the left and wealthiest to the right. Wealth is shown by monthly income.

About: explains the purpose of Dollar Street, with overview videos.

Use these tools to search by family home item, continent or country.

Select a family and read the description, including their hopes for the future. Click on a photo to explore everyday items in the home.

Map: shows the location of all the families on a world map. Click on the location to visit.

Imagine the world as a long street, Dollar Street, poorest to left and richest to right, everybody else in between. The house numbers equal the income you have. We sent a team of photographers across the world to visit families in their home and take photographs according to 135 categories. We want to show how people really live. It seemed natural to use photos as data so people can see for themselves what life looks like on different income levels.

D **Anna Rosling Rönnlund, the creator of Dollar Street**

Rosling Hans (2018) *Factfulness* page 44:

'Your most important challenge in developing a fact-based world view is to realise that most of your experience of life is from Level 4, where you live ... When in Level 4 everyone in Levels 1–3 can look equally poor ... even a person on Level 4 can look poor. '

So we see the world as a gap between rich and poor. To help you overcome this 'Gap Instinct', the Gapminder Foundation has also developed Dollar Street, C, created by Anna Rosling Rönnlund (D). In this lesson, you will use this website to start to develop a fact-based view of the world.

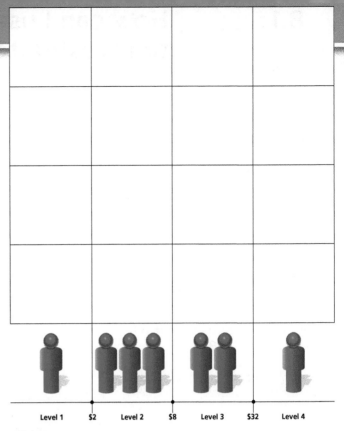

E **Dollar Street research recording sheet**

Activities

1 Look at the cover of the book *Factfulness* (A).
   a) Who are the authors of the book?
   b) What is the book trying to achieve?
   c) Write a sentence to explain what the book means by the term 'Gap Instinct'.
2 Look carefully at diagram B, used in *Factfulness*.
   a) What are the categories for the four income levels?
   b) How is the world's population of 7 billion divided up across the income levels?
   c) What percentage of the world's population lived at Level 1 200 years ago?
3 Use screenshot C, quote D and the Dollar Street website for the following activities.
   a) What is Dollar Street?
   b) Go to the website, and watch the 'Quick tour'. Write three sentences to explain how the website is organised.
   c) Select a family at income Level 1, and explore their homes and everyday items.
   d) Write a paragraph to describe their life.
   e) Identify ways that your home is similar and different from theirs. Summarise your points in a table.

4 Make a large copy of the grid shown in E on a sheet of A4 paper to use as a recording sheet for research on the Dollar Street website.
   a) Consider your investigation of Dollar Street in Question 3, and identify four everyday items, to compare across the income groups, that will give you a good idea of how life is different across the groups.
   b) Add a column on the left-hand side of the table to add a title for each item selected.
   c) Add a row at the top of the table to add the name of the family, their income, their country and their hopes for the future.
   d) Select a family for each income group on the website, and record your findings for each on your recording sheet.
   e) Use your table to help you write a paragraph to describe how life is different at each of the four income levels.
5 Conclude your understanding of a fact-based view of the world.
   a) Write a sentence to explain why we tend to have a 'Gap Instinct', from our life at Level 4.
   b) Write a paragraph to explain how this enquiry has helped you to develop a fact-based view of the world.

Learning objectives

▶ To use a dot map to investigate population distribution patterns.

▶ To calculate population density.

▶ To compare a dot map with a cartogram.

The number of people living in a particular place is called the population. World population is constantly changing, and geographers can use a variety of geographical data to help them visualise and understand these changes. In the next three lessons, you will explore population distribution and change at a variety of scales.

- Population distribution is the pattern of where people live and how populations are spread.
- Population density is the number of people living in a given area, usually a square kilometre.
- Population density per square kilometre is calculated as follows:

$$\frac{\text{Total population}}{\text{Area}}$$

A Population and area for ten countries, 2018

Country	Population, 2018	Area, km²
Russia	145,734,038	17,098,246
Bangladesh	161,376,708	147,570
China	1,411,582,266	9,706,961
Singapore	5,757,499	710
Mexico	126,190,788	1,964,375
Bahrain	1,569,446	765
UK	67,141,684	242,900
UAE	9,630,959	83,600
Nigeria	195,874,683	923,768
Brazil	211,049,527	8,515,767

Dot maps

Population density can be plotted as a dot map. Each dot is the same size and represents a certain value. Map B shows a dot map for the world's population. Each dot represents 50,000 people. The map clearly shows that the world's population is not evenly distributed.

A dot map is useful for identifying overall patterns, but it lacks accuracy. In South East Asia, for example, the areas are so densely populated that the dots are too close together to count. In areas where there are few people – sparsely populated areas, for example the Sahara Desert, or eastern Russia – there may be a single dot. The people in these areas, however, are in reality not located where the dot is on the map, but are spread over a very large area. Population distribution can also be presented as a cartogram (see C).

Global Population Density

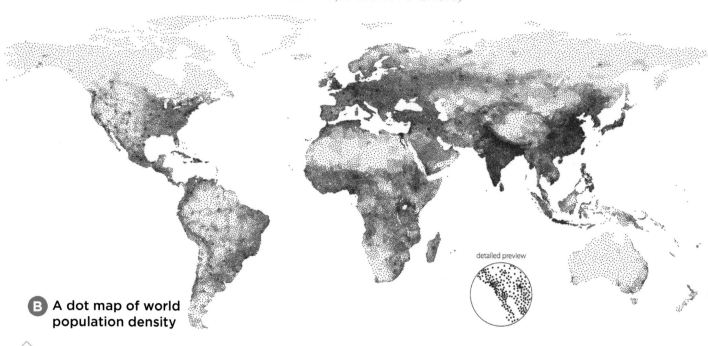

detailed preview

B A dot map of world population density

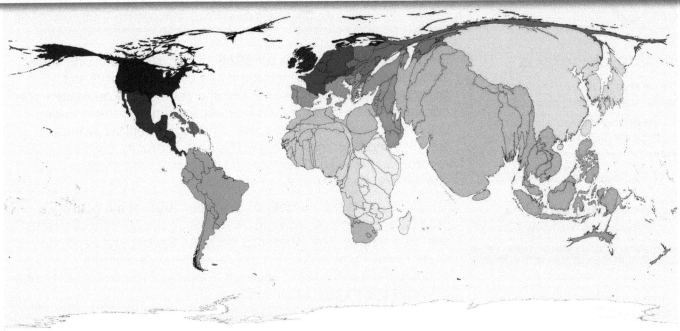

C Cartogram showing world population, 2018, Worldmapper

Activities

1 What is population density, and how is it measured?

2 Make a copy of table A.
   a) Calculate the population density for each country. Add a column to your table to record the density.
   b) Complete the table by adding your calculations and the rank order of each country according to their population density. Add a column to your table to record rank order.
   c) What do you notice about the countries with the highest density compared to those with the lowest?

3 Look carefully at map B.
   a) What is a dot map?
   b) Write a paragraph to describe the world's distribution of population, naming five sparsely and five densely populated regions as evidence of the distribution pattern. Use map B on Lesson 1.3 to help you.
   c) Identify an advantage and disadvantage of using a dot map to show population distribution.

4 Look carefully at map C.
   a) What type of map is this, and what does it show?

   b) Which countries and continents are presented smaller and larger than their area?
   c) Write three sentences to describe what this map shows you about population distribution.

5 Compare B and C.
   a) How is the data presented in map C different from the data presented in map B?
   b) Write two sentences to explain which map you find more useful for identifying the world's population distribution.
   c) Identify an advantage and a disadvantage of the cartogram.

6 The following websites show other ways of presenting the world's population.
   http://luminocity3d.org/WorldPopDen/
   https://pudding.cool/2018/10/city_3d/
   a) Go to each website.
   b) In each case, describe how population distribution is shown.
   c) Explain which website provides the more helpful way of presenting population distribution.

7 Write a conclusion for this lesson identifying what you think is the best way of presenting population distribution, giving two reasons to justify your choice.

How can I use statistics, line graphs and a model to investigate population change?

Learning objectives

▶ To be able to draw a line graph to show population change.

▶ To understand population change within the Demographic Transition Model.

A World population data, https://ourworldindata.org/

Year	Population
10,000 BC	4 million
0	250 million
1700	600 million
1800	1 billion
1900	1.65 billion
1930	2 billion
1960	3 billion
1974	4 billion
1987	5 billion
1999	6 billion
2011	7 billion
2024	8 billion
2039	9 billion
2056	10 billion
2088	11 billion

Demography is defined as the study of human populations: their size, composition and distribution, as well as the causes and consequences of changes in these characteristics. Populations are constantly changing; they grow or decline through the changes in birth, death and migration rates. Demographers study these statistics and how they change. They use censuses, birth and death records, and surveys. They search for explanations of demographic change and their implications for societies.

The **birth rate** is the number of births per 1000 of the country's population, each year, and the **death rate** is the number of deaths per 1000 people. The difference between the birth rate and the death rate is the **natural increase or decrease** in a country's population. These rates change for a variety of reasons as a country develops and its population evolves.

Table A shows how the world's total population has changed through time. The Demographic Transition Model, C, was developed in the mid-twentieth century to explain this change. The model is based on the observation of similar population growth patterns in mainly European countries, over 200 years, as their economies developed. Originally identifying four stages, a fifth stage was added towards the end of the century, and some demographers suggest there may be a sixth stage. We are still not sure how this model will evolve. Rapidly developing countries in Asia and Africa may not follow the same population changes experienced more slowly by European countries.

5 Millions of people are starving in the country due to a succession of famines, leading to crop failure and lack of clean water.

1 Young girls in the country now have access to education. Eventually educated mothers decide to have fewer children, in order to pursue a career.

2 A typhoid epidemic has led to the deaths of many young children.

6 People predominantly live in rural areas, grow crops and search for water to survive. They have lots of children to help on the land, in the hope that some survive.

3 The government needs to deal with economic problems when there are not enough working-age people, as the country has increasingly high numbers of retired, elderly people.

7 Over the last 50 years, a network of hospitals, sewage systems and refuse collection has been developed across the country.

4 The government of the country has decided to attempt to control population growth by offering families financial incentives to have fewer children, as well as access to free contraception.

8 People have moved to cities and have jobs in the industries and services that have developed there. They have access to schools, hospitals and clean water supplies.

B Factors that contribute to population change

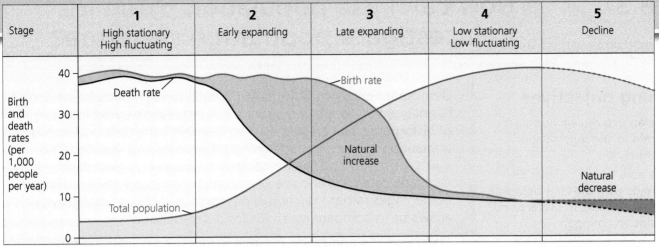

Stage	1 High stationary High fluctuating	2 Early expanding	3 Late expanding	4 Low stationary Low fluctuating	5 Decline
Examples	UK before 1780 Niger today	UK, 1780–1880 Afghanistan today	UK, 1880–1940 Brazil today	UK today	Russia today
Birth rate	High				
Death rate	High				
Population change	Gradual increase				
Factors affecting a country					

C **The Demographic Transition Model (DTM)**

Activities

1 What do demographers study?
2 Look carefully at the population data shown in A.
   a) Draw a line graph to show how the world's population has changed from 10,000 BC to 2088. Look back at Lesson 1.10 to help you. Use a scale of 1 cm = 1000 years for the x-axis, and 1 cm = 1 billion people for the y-axis.
   b) 2024–2088 are estimates made by the United Nations. On your graph, use a dashed line to show this.
   c) Add a title to the graph and label each axis.
   d) It took 12,000 years for the world's population to reach a billion. Calculate how many years it took for the population to grow by each billion after that. Add these calculations to your graph.
   e) Watch the following video:
   www.gapminder.org/answers/how-did-the-world-population-change/
   f) From the video, make a list of the following factors and annotate your graph with them:

   - factors that meant the population did not grow quickly before 1800
   - factors that led to world population growth
   - factors that explain why population growth is estimated to slow down.

   g) Write a paragraph to describe and explain how the world's population has changed.
3 **a)** What is the Demographic Transition Model?
   b) Why do you think demographers created the model?
4 Look carefully at the model C and factors B.
   a) Make a copy of the model and table in C.
   b) Complete the descriptions of the birth and death rates and population change shown by the model for stages 2–5.
   c) Consider which stage of the model each statement (1–8 in B) best fits, and write it into the correct 'Factors' row of your table.
   d) Write a paragraph to describe how the population of a country changes at each stage of the model. Use the graphs and explain these changes by justifying the choices you made in Question 4c).

How can I use population pyramids to investigate population structure?

Learning objectives

▶ To be able to draw a population pyramid to show population structure.

▶ To be able to use population pyramids with the Demographic Transition Model to understand population change.

Demographers use graphs called population pyramids (see A and C), to analyse the population structure. A population pyramid is a pair of histograms, placed side by side, that show the distribution of a population for different age groups, for males and females. The x-axis is used to plot the percentage of a country's population in each age category, and the y-axis lists the age categories. Using percentages rather than actual numbers in each age category allows us to compare pyramids for different countries.

The shape of a population pyramid can be used to interpret a population. For example, a pyramid with a very wide base and a narrow top section suggests a population with both high birth and high death rates. However, a pyramid with a wider top half and a narrower base would suggest an ageing population with low birth and death rates. These graphs are very useful to governments to better understand the country's population structure, in order to plan for the future.

A How to draw a population pyramid

B Data for population pyramid for Japan

Step 1: Draw the y-axis for each age group of five years.

Step 6: Add a title for the graph, axes and the male and female sides of the graph.

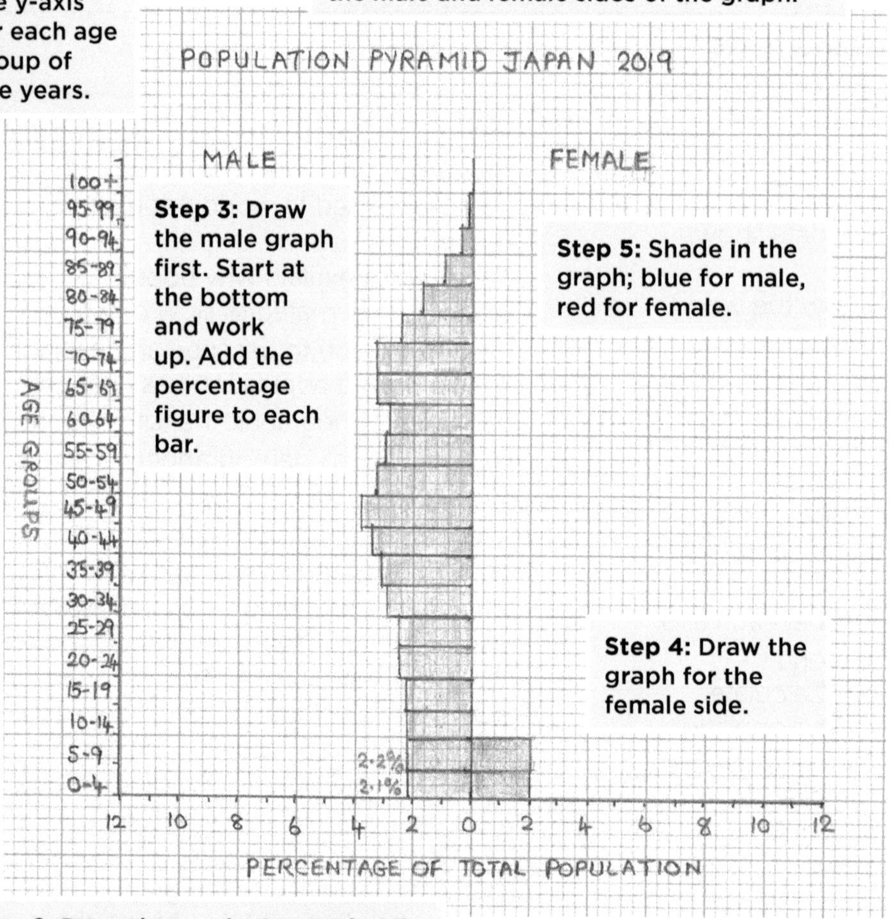

POPULATION PYRAMID JAPAN 2019

MALE FEMALE

Step 3: Draw the male graph first. Start at the bottom and work up. Add the percentage figure to each bar.

Step 5: Shade in the graph; blue for male, red for female.

Step 4: Draw the graph for the female side.

AGE GROUPS

PERCENTAGE OF TOTAL POPULATION

Step 2: Draw the x-axis. Remember 0 is in the middle, with an axis for males to the left and for females to the right.

Age group, years	Percentage of the total population, 2019	
	Males	Females
100+	0	0.1
95–99	0.1	0.3
90–94	0.4	1
85–89	1	1.8
80–84	1.7	2.4
75–79	2.4	3
70–74	3.3	3.8
65–69	3.3	3.5
60–64	2.9	3
55–59	3	3.1
50–54	3.3	3.3
45–49	3.9	3.8
40–44	3.5	3.4
35–39	3.1	3
30–34	2.9	2.7
25–29	2.5	2.4
20–24	2.5	2.3
15–19	2.3	2.2
10–14	2.2	2.1
5–9	2.2	2.1
0–4	2.1	2

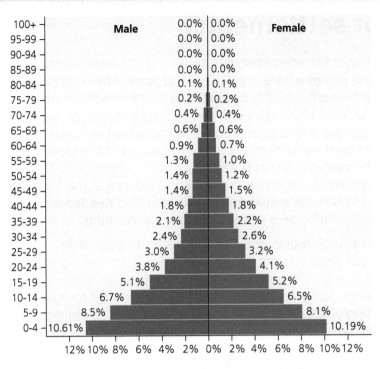

	Male	0.0%	0.0%	Female	

Sketch population pyramids

Sketch pyramids (see D) can be drawn to summarise the characteristic features of the population structure for each stage of the Demographic Transition Model.

C Population pyramid for Niger, 2019

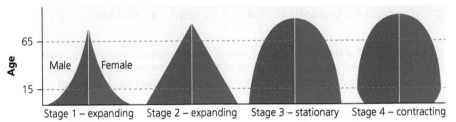

D Sketch pyramids

Activities

1 a) What is a population pyramid and what data is it used to show?

b) Why are percentages in each age category shown rather than actual numbers of people?

2 Look carefully at population pyramid A.

a) Make a copy and complete the pyramid on graph paper, using the data in table B.

b) Calculate the percentage of young people in Japan, aged 0–19, and the percentage of old people, aged 65+. Label these percentages on your pyramid.

c) The remaining categories are the working population called the 'economically active', aged 20–64. Calculate this percentage and label it on the graph.

3 Compare your population pyramid for Japan with the pyramid for Niger, C.

a) Write a paragraph describing the distribution of young, economically active and old age groups for the two countries.

b) Which country has a high birth rate? Justify your answer.

c) Which country has low birth and death rates? Justify your answer.

4 Look carefully at the four sketch pyramids in D.

a) Make copies of each pyramid in the appropriate stage on your Demographic Transition Model, drawn in Lesson 8.2, Question 4.

b) Annotate each of your pyramids to explain how its shape represents each stage of the model.

c) Your pyramid for Japan has a different shape to the four sketches, as Japan is in stage 5 of the model. Draw a sketch of this shape on your model diagram.

How can I draw sketch maps from OS maps to investigate the site and situation of settlements?

Learning objectives

▶ To be able to draw annotated sketch maps from OS maps.

▶ To interpret OS maps and aerial photos to identify location factors for settlement.

▶ To understand the different site and situation factors for settlements.

Settlements are places where people live and work. The physical land on which a settlement is built is called the site. Many settlements in the UK are at least 1000 years old, so we need to consider the needs of the people at that time. People had to grow their own food, find their own water supply, use the resources of the local area and defend themselves against potentially hostile neighbours. The site also had to provide land that was well drained, not at risk from flooding and sheltered from extreme weather. It also needed a good view of the surrounding area to see enemies approaching.

Gradually, some settlements grew into towns and cities, while others remained small villages. Those that grew often had a good situation. The situation of a settlement is its location, in relation to the surrounding area. Some settlements were at the focus of routeways, often where two valleys met at the confluence of rivers or bridging point of a river, where it narrowed to allow the span of a bridge. As trade and industry developed in the UK, during the Industrial Revolution, some settlements developed into large cities and ports, as goods could be brought in and out by road, rail and ship.

A How to draw a sketch map from an OS map

Step 1: Using a pencil and ruler, draw a frame, to the same scale as the area on the OS map you are sketching.

Step 2: Using a pencil and ruler, draw the grid squares as they appear on the OS map. Add grid references for the corners of your frame. These squares will act as guidelines when you start to draw your sketch.

Step 5: Label and annotate the key features on your sketch map, in this case, the site and situation of the settlement.

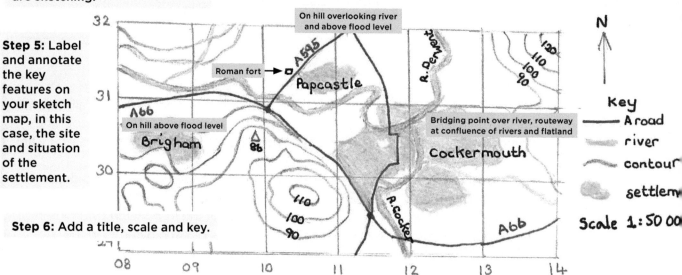

Step 6: Add a title, scale and key.

Step 4: Once you are happy with the sketch, use coloured pencils to make the features stand out – for example, shading a river blue.

Step 3: Use a pencil to carefully draw the key features. Start with an outline of these features, positioning them using the grid squares. Draw physical and human features.

Sketch map A is a simple map that shows some but not all of the detail of an OS map. Sketch maps can be used to show the relationship between physical and human features. The gridlines on the OS map make it possible to draw a fairly accurate sketch map as they act as guidelines for drawing the information.

B Aerial photograph of Corfe Castle

Activities

1 Write definitions of settlement, site and situation.

2 Why did early settlers have to choose the site of a settlement carefully?

3 Why did some settlements grow into towns and cities?

4 Look carefully at sketch map A and compare it with map-flap E.
   a) Locate the grid square frame for the sketch map.
   b) Write a paragraph to describe the sites of Brigham, Papcastle and Cockermouth. Include map evidence with grid references in your description.
   c) Why do you think Cockermouth grew larger than the other two settlements?

5 Look carefully at aerial photo B and compare it with map-flap D.
   a) In which direction was the camera pointing when the photo was taken?
   b) Name features 1–5 marked on the photo and locate them on the map, giving the six-figure grid references for each.

6 Draw a sketch map to show the site and situation of the village at Corfe Castle. Use the guidance provided in A, map-flap D and photo B to help you. Annotate the sketch with the site and situation location factors.

7 Look back at Lesson 1.5 or 1.13 and draw a sketch map to show the site and situation factors for either Yarm or Scarborough.

8 Draw a sketch map to show site and situation factors for a settlement in your local area.

How can I identify coastal erosion landforms using photos, drone videos and OS maps?

Learning objectives

▶ To be able to identify coastal erosion landforms on OS maps.

▶ To compare OS maps with photos and drone videos.

▶ To be able to follow a route and measure its distance.

Old Harry Rocks, photo A, is three chalk landforms, located at Handfast Point, on the eastern edge of the Isle of Purbeck, part of the Jurassic Coast, a UNESCO World Heritage Site. The headland is a part of Ballard Down, owned by the National Trust. This coastline, including Old Harry Rocks, was created through thousands of years of erosion by the sea.

A Old Harry Rocks, Dorset, looking north

Coastal erosion landforms

- Waves continually hitting the resistant chalk rocks bang loose rocks into weaknesses, joints and faults in the chalk cliffs. These small weaknesses are gradually enlarged. Over thousands of years, the waves create small bays with promontories, giving the coast a jagged appearance (see aerial photo B).

- Caves are eroded on either side of the promontories and eventually merge, forming an arch. In the past, there were arches between rock formations 2 and 3 in photo A. These were widened by erosion, until the roof of the arches collapsed, through lack of support, creating two stacks (pillars of chalk), no longer connected to the main cliff.

- Gradually the stack wears to leave a stump.

- The name Old Harry Rocks actually refers only to the single stack of chalk which stands furthest out to sea. Until 1896, there was another stack beside this known as Old Harry's Wife, but erosion caused it to tumble into the sea, leaving just a stump.

- Eventually, the cliff is reduced to a wave-cut platform – a platform of flat rock created by waves where the cliffs have been eroded away, leaving a flat rock remnant of where the cliff used to be, at beach level. Look at the OS map legend in the Water Features section for the symbol for flat rock.

Drones

Drones are increasingly used to survey and explore landscapes which are often difficult to access. The cliffs at Old Harry are high and steep-sided, making it difficult to see the full range of coastal erosion and assess the potential for future cliff fall. Regular drone flights over the area allow digital reports to be created, which can monitor changes in the cliff remotely.

Activities

1 Old Harry Rocks and Ballard Down are owned by the National Trust, which protects this beautiful and important landscape. The cliffs are very dangerous, so the Trust maintains a trail footpath for visitors. Look at map-flap D. The route of the trail begins at the pub car park at grid reference 038826, following the coastal footpath east to Old Harry, then south-west towards Ballard Point 048815, and then west to 030811. Here the trail changes direction down the slope in a north-easterly direction back to the pub car park.
   a) Follow the route on map-flap D.
   b) Measure the distance along the route. Look back at Lesson 1.7 to help you.
2 Look carefully at photo A.
   a) Draw a fieldsketch from the photo. Look back to Lesson 1.13 to help you.

   b) Label coastal landforms 1–4.
   c) Annotate your sketch to explain how the coastal erosion landforms are created.
3 Compare aerial photo B with map-flap D.
   a) In which direction was the camera pointing when this photo was taken?
   b) Name the physical and human features labelled 1–10, and give the six-figure grid reference for each.
4 Conduct a video search on the internet: 'drone flight Old Harry Rocks'.
   a) Watch the video and compare it with map-flap D, to follow the route of the drone.
   b) Identify three ways this drone flight improves your understanding of the area.
   c) Write a paragraph to explain the advantages of drones for geographers.

How can I use photos, OS maps and sketch maps to identify the impact of geology on shaping a coast?

Learning objectives

▶ To be able to identify landforms on OS maps.

▶ To compare OS maps with photos and geology maps.

▶ To be able to draw a sketch map and use it to explain the impact of rock type on a coastal area.

The geology (or rock type) that makes up an area has a major impact on how the sea erodes and shapes the coast. Soft rocks are more easily eroded by the sea than hard rocks. Along the east coast of the Isle of Purbeck, there are alternate layers of hard and soft sedimentary rocks (see B). These rock layers run at right angles to the coast, leading to the formation of headlands and bays. This is because the sea can erode the soft rocks more quickly than the hard rocks. The chalk rocks that outcrop at the coast at Ballard Down are not horizontal; they have been folded, or buckled up, by the movement of rocks over millions of years, so they dip. A rising angle of rock at the coast can lead to a very steep and high cliff profile, see aerial A. In this lesson, you will use an OS map, geology map and aerial photo to assess the impact of geology on the shape of this coastline.

(A) Aerial photo of Old Harry Rocks and Ballard Down, Dorset

B The geology of the East Dorset coast

Before and during erosion

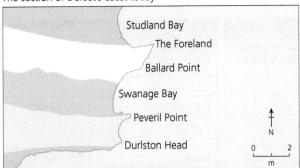

Bagshot beds (soft)

Chalk (hard)

Wealdon clays and greensands (soft)

Key

Purbeck and portland limestone (hard)

Easy to erode
Hard to erode

The section of Dorset's coast today

Studland Bay
The Foreland
Ballard Point
Swanage Bay
Peveril Point
Durlston Head

N

0   2
m

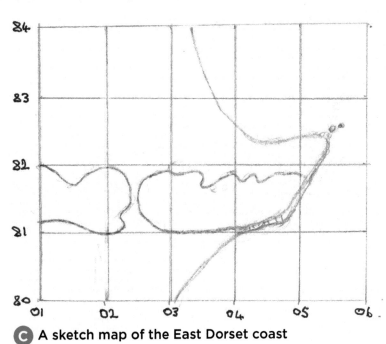

84

83

82

81

80

01   02   03   04   05   06

C A sketch map of the East Dorset coast

Activities

1 Compare aerial photo A with map-flap D.
   a) In which direction was the camera pointing when the photo was taken?
   b) Identify features 1–10 labelled on photo A.
   c) Look closely at the cliffs from location 3 to 1 on the photo, and write a sentence to describe how the cliff changes between these points.

2 Draw a cross-section for Ballard Down from 035810 to 035823 (use Lesson 6.1 for guidance).

3 Compare your cross-section with photo A, and write three sentences to explain your answer to Question 1c).

4 Look carefully at sketch map C. Find and compare it with map-flap D.
   a) Make a copy of the sketch map.
   b) Add the contour pattern for Ballard Down.
   c) Label features 1–9 from Question 1b) on your sketch map.
   d) Compare your sketch with geology map B. Add and shade the different rock types on your sketch map.
   e) Label the headlands and bays on your sketch map.
   f) Annotate your sketch map to explain how the geology and the dip in the chalk have led to the sea creating this shape of coast.

5 Write a paragraph, using all the evidence you have collected in this lesson, to explain the impact of rock type on the shape of the East Dorset coast.

Enquiry skills: How can I use geographical data to make a decision about coastal management?

Learning objectives

▶ To identify a coastal erosion problem using an OS map and photos.

▶ To be able to consider different options for a coastal management scheme.

▶ To be able to consider different points of view and justify a decision about coastal management.

In this lesson, you will analyse a range of quantitative and qualitative data to decide upon a coastal management plan along a stretch of coast in Scarborough, North Yorkshire. You will use some of the data considered by Scarborough Borough Council in their own decision-making process.

A **Costs that would result from using the 'Do Nothing' strategy, Scarborough Borough Council**

Category	Estimated cost of 'Do Nothing' plan	Category	Estimated cost of 'Do Nothing' plan
Spa Complex	£91,174,000	Harbour disruption and dredging	£320,000
Residential properties	£11,860,000	Cliff lift	£826,000
Non- residential properties	£1,844,000	Risk to life	£52,687,000
Tourism	£12,784,000	**Total**	£171,496,000

The problem: threat to the Spa Complex from coastal erosion

The context

Scarborough is a popular coastal resort in North Yorkshire, attracting over 3.5 million visitors a year. The Spa Complex is an iconic Grade II listed building and of great historical significance to the town. It is a major visitor venue as a conference and entertainment centre, hosting festivals, theatre productions, concerts and other events. On top of the cliff, above the Spa, are large numbers of Regency and Victorian properties, dominated by hotels, offering panoramic views of the castle and coastline.

Reasons for concern

The coastline has a history of cliff landslides. On 3–5 June 1993, a landslide occurred at the Holbeck Hotel. The hotel was destroyed and the cliff was cut back, creating a 200 m wide semi-circular promontory on the beach.

The cliff behind the Spa is only 0.5 km away from the Holbeck slope. It has the same geology, with the same risk of landslide.

Factors that led to the 1993 landslide

A period of heavy rainfall; drainage issues; the geology of the cliff – unconsolidated glacial till resting on bedrock of mudstone and sandstone.

Action

There are three main options to prevent the threat to the Spa Complex from coastal erosion – see C.

B Scarborough South Bay from 049890

Site of Holbeck landslide

① ②

Option 1: Do nothing

- No further maintenance or intervention at seawall and cliffs.

Results of this approach:

- Eventual deterioration and collapse of existing seawall.
- Increased damage from wave over-topping seawall.
- Increased slope failure, potentially impacting properties on the cliff top, including 380 households and 38 commercial properties, including a number of hotels.
- Risk of damage or loss of the Spa Complex.
- Risk to public safety, potential fatalities with landslide.

Cost: £171,496,000 – see table C.

Option 2: Do minimum

- Maintain the existing structures and repair shallow landslips and seawall, as is done now.

Results of this approach:

- This would only delay the onset of the 'Do Nothing' strategy.

Cost:

- £78,000 a year to maintain seawall further.
- £16,000 a year maintaining the cliff.
- Plus costs outlined in table A – £171,496,000 – when collapse occurs.

Option 3: Do something to hold the line

- Rock armour in front of the seawall, in combination with a wave wall to reduce overtopping, with cliff stabilisation.

Result of this approach:

- Provides more stability to the defences that are already in place to prevent the failure of the seawall. The wave wall will reduce the risk of over-topping.
- The cliff stabilisation scheme along with the work on the seawall will stop the cliff from moving.
- Damage to buildings at the top and base of the cliff would be avoided.
- Protects the tourist economy of the town.
- Reduces the risk of loss of life.

Cost: £16 million.

C Options for a solution, Scarborough Borough Council

Public consultation, 2015

D Stewart Rowe, Principal Coastal Officer, Scarborough Borough Council

There is already evidence of cliff and shoreline instability. We monitor it all the time and we are seeing evidence of movement not dissimilar to when the Holbeck went. It could be now, it could be years ahead, but it will happen. The difference to Holbeck is people saw the gardens starting to move and had time to evacuate. What if there is a full concert at the Spa and it goes and there is no time to evacuate the properties behind? We have a risk and we need to manage it.

E Artist's impression of rock armour in front of Spa seawall as part of Option 3

The existing Victorian stone seawall is a key part of the historic heritage of the Spa. The wall complements a building of great architectural merit. Your solution of a heap of rubble stacked against the wall is an insult to the Victorian engineers. The Sons of Neptune strongly object to the scheme, we want to see the existing wall restored, as part of Scarborough's heritage.

F Freddie Drabble, Sons of Neptune, local environmental group

Activities

1. Compare photo B with OS map A of Scarborough in Lesson 1.13.
   a) Identify and name the place where photo B was taken from, and the direction in which the camera was pointing.
   b) Name and give the six-figure grid references for features 1 and 2 in photo B.
2. Go to Google Earth and search 'Scarborough UK'. Explore the area shown in photo B, using the aerial view and street view. Write a paragraph to describe why you think the Spa Complex and coastline is popular with tourists.
3. Read the options facing Scarborough at the public consultation meeting in 2015, C.
   a) Look at table A. Calculate the damage values into percentages.
   b) Draw a pie chart to present the data.
   c) Write five sentences to describe what the chart shows and why this is a problem for the council.
   d) Option 3 partly involves placing rock armour in front of the existing seawall. Write a paragraph to describe points for and against this aspect of Option 3.
   e) With a partner, discuss the advantages and disadvantages of each option, and complete a two-column table summarising them.
   f) Conduct a class public meeting to discuss each option.
4. **Decision-making**: Decide which option you think is best for the future of this coastline. Write a paragraph to justify your decision.

Enquiry skills: How can I annotate an aerial photo to consider a decision about coastal management?

Learning objectives

▶ To be able to consider a decision about a coastal management scheme.

▶ To be able to draw a sketch of an aerial photo.

▶ To be able to interpret and annotate an aerial photo to describe a coastal management scheme.

In Lesson 9.3, you discussed three options for coastal management at Scarborough Spa, to reach your decision about what to do. Scarborough Borough Council had a similar discussion to you to reach a decision about what to do to protect the coast at Scarborough Spa. They decided on Option 3, but as a phased plan (see quote A). Work on the first phase of the Spa Slope Stabilisation Scheme began in May 2018 and was due for completion in December 2019. This work was estimated to cost £14.7 million – £11.619 million funded by the Environment Agency and £1.212 million from North Yorkshire County Council, with the rest funded by Scarborough Borough Council, who took out a loan of £1.88 million.

There were a number of reasons why the council decided not to implement all of Option 3. A number of people campaigned against placing rock armour in front of the Spa seawall, but ultimately the council could not afford the preferred Option 3; they would have needed to borrow an additional £10.8 million to do this. As a result, they opted for this phased approach, dealing with the major risk of landslides behind the Spa first (see B). In this lesson, you will draw a sketch from an aerial photograph of the area to be managed (see C), and annotate the key elements of the scheme.

> The preferred option will be delivered in two phases. The first phase involves stabilising the slope along the rear of the Spa complex from the cliff railway, relaying of cliff access paths and landscaping/replanting together with ongoing inspection and maintenance of the seawall. The timing of the second phase, over the next 50 years, will depend on the rate of climate change and the rate of deterioration of the seawall. It will consist of a rock revetment, placed in front of the seawall.

A **Stewart Rowe, Principal Coastal Officer, Scarborough Borough Council**

Erosion prevention mesh · Geocell profile · Nail head plate · Articulating boss · Domed nut or hex nut and hemisphere

Once the soil nails were installed, a large plate and bolt were screwed onto the end to hold it in place. Soil was then fixed over the nails and grass seeded on the slope. New paths, fencing and landscaping of the slope was then carried out.

Shallow slip

Soil nails, 3 metres long, drilled into cliff face to reinforce shallow slip.

Piles drilled 18 to 38 metres through the cliff and into solid bedrock below, to reinforce deep slip.

Deep slip

Spa Building

Seawall – ongoing maintenance

Sea

Sandstone bedrock

Piles drilled up to 3 metres into bedrock

B **Phase 1: Scarborough Spa Slope Stabilisation Scheme**

C Aerial photo of Scarborough Spa before stabilisation

Activities

1 Conduct a search on the internet for Holbeck Hall landslide Scarborough. Try to find a news report with a video of the event.
 a) Write three sentences to explain what Scarborough Borough Council learnt from the Holbeck landslide of 1993.
 b) Write two sentences to explain why you think this experience motivated the council to protect the cliff behind the Spa.

2 Read quote A carefully.
 a) Write a paragraph to describe the coastal management decision made by the council.
 b) Write three sentences to explain why the council decided on two phases to the scheme.

3 Look carefully at aerial photo C.
 a) Draw your own sketch of photo C.
 b) Colour your sketch and add a title.
 c) Match the following labels to the numbers 1–6 on photo C: seawall, cliff railway, buildings on the Esplanade at the cliff top, Spa buildings, beach, cliff to be stabilised. Then add the labels to the correct places on your sketch.

4 Look carefully at diagram B and re-read Lesson 9.3.
 a) Annotate your sketch to explain the problems facing this coastline.
 b) Annotate your sketch to show how the cliff will be stabilised in phase 1 of the scheme.

5 What factors will determine when Scarborough Borough Council will need to implement phase 2 of the scheme?

Enquiry skills: How can we conduct a whole-class enquiry into the diversity of Asia?

Learning objectives

▶ To compare physical and population atlas maps of Asia.

▶ To locate and describe places using latitude and longitude.

▶ To compare levels of development and ways of life in Asia.

Asia is the world's largest continent, with 30 per cent of the total land mass and over 60 per cent of the world's population. It is made up of 48 countries. Asia is a very diverse continent, this means it has a great variety of physical and human geography. In this lesson, you will conduct a geographical enquiry to capture just a glimpse of this diversity.

Activities

1 Look carefully at map A.
   a) Name the continent that borders the western side of Asia, and what forms this border.
   b) Name the seas and oceans that border Asia.

2 a) Locate the landscapes at the following coordinates on maps A and B:
   i) 30°N 90°E
   ii) 24°N 90°E
   iii) 35°N 137°E
   iv) 23°N 113°E
   v) 29°N 77°E
   vi) 12°N 124°E
   vii) 27°N 71°E
   viii) 30°N 82°E
   ix) 5°N 118°E
   x) 56°N 160°E

   b) Draw a table with columns to record each coordinate, the country and a description of the relative location.
   c) Mark and name each location on an outline map of Asia.

A Physical map of Asia, © *Philip's Essential School Atlas*

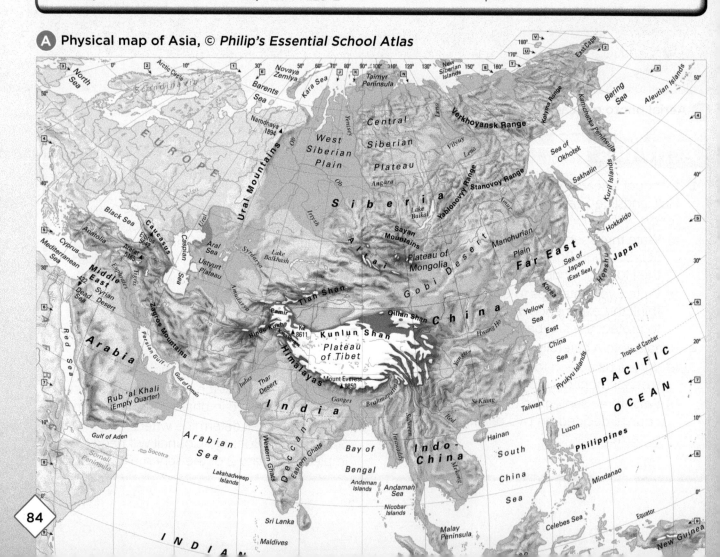

POPULATION DENSITY

Inhabitants per km²

	Over 200
	100 – 200
	50 – 100
	10 – 50
	1 – 10
	Under 1

Population of major cities in millions

	Over 10
	5 – 10
	2.5 – 5
	1 – 2.5
	0.5 – 1

B Population density map for Asia, © *Philip's Modern School Atlas*

Activities

3 **a)** Conduct a web enquiry with a partner. Search two of the locations from Question 2a on the internet using http://confluence.org/ and Google Earth. Look carefully at the photos, descriptions, satellite images and Google Street View for your locations.

b) Write a paragraph for each of your chosen locations to describe what the place is like using as many of the enquiry questions introduced in Lesson 1.1 as you can. Include a title for each paragraph that gives the place name of the location, the country and the region it is in.

c) Share what you have discovered about your two locations with the rest of the class. Discuss your findings and summarise what you have discovered about the diversity of Asia.

4 Look carefully at map B.

a) Write a paragraph describing the distribution of population in Asia.

b) Compare maps A and B and think about what you discovered about the continent in Question 3. Write a paragraph to explain the reasons for the population distribution across Asia.

5 Look back at map A in Lesson 7.1. It shows the global distribution of GNI per capita.

Draw a two-column table. In the first column, write the names of the following ten Asian countries: Bangladesh, Cambodia, China, India, Indonesia, Japan, Kazakhstan, Myanmar, Nepal, South Korea.

In the second column, use the map to record which GNI per capita category each country is in.

6 Go to the Dollar Street website: www.gapminder.org/dollar-street.

a) Select the search tool in the left-hand corner, click on the world box, select 'Asia'. The Dollar Street home page will change to show families in four income groups across Asia.

b) Visit families across the four income groups.

c) Write a paragraph to describe their life, and how it is similar and different from yours.

7 Conclusion: Reflect on what you have discovered about Asia, and summarise ways it is diverse.

Learning objectives

▶ To understand how geographers can use satellite imagery.

▶ To interpret satellite images to identify how places change.

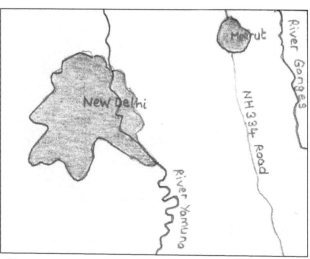

Ⓐ **Sketch map of New Delhi**

Remote sensing is the science of obtaining information without physically being there, using satellites. These satellites orbit 400 miles above the Earth, scanning the Earth's physical features. The Landsat series of satellites has been recording information since the first launch in 1972, providing live data with every orbit of the planet. The signals from the satellites are converted into satellite images using computers. Remote sensing is particularly helpful in understanding how Earth is changing over time, by comparing images, such as B and C which are examples of how remote sensing can be used to track the growth of cities and changes to the environment over several years. Satellite instruments can now also measure vegetation coverage, sea-ice fluctuation, sea level, sea surface temperatures and concentrations of atmospheric gases.

New Delhi

New Delhi, India's capital city, is situated in North India at 28.61°N 77.23°E, on the Yamuna River, 648 miles from the border with Nepal. It is one of the fastest growing cities in the world, with a population of 27 million. The United Nations forecasts that by 2028 the population could outstrip Tokyo to make it the world's biggest megacity. Satellite images B and C clearly show how the city has grown.

Activities

1 **a)** What is remote sensing?
   b) What is Landsat?
2 Compare satellite images B and C with sketch map A.
   a) Name features 1–5.
   b) Compare the two satellite images and write three sentences to describe how New Delhi changed between 1989 and 2018.
3 Search New Delhi on Google Earth.
   a) Zoom in and out of the area shown in A, B and C.
   b) Use the enquiry questions from Lesson 1.1 to write a paragraph describing what the area is like.
4 Go to the Google Earth Timelapse website: https://earthengine.google.com/timelapse/

Timelapse is a zoomable video that uses sequences of satellite images of places over the past 35 years to animate change.
   a) Type 'New Delhi' into the search window. Watch the animated satellite images of the city change and write two sentences to describe what they show.
   b) Scroll down the examples on the left-hand panel to the urban growth time lapses for 'Dalian, China' and 'Naypyitaw, Myanmar', and watch the animations.
   c) Write five sentences to describe how all three cities are changing.
5 Reflect on what you have discovered in this lesson about using satellite images. Write a paragraph to explain why you think satellite images are useful to geographers.

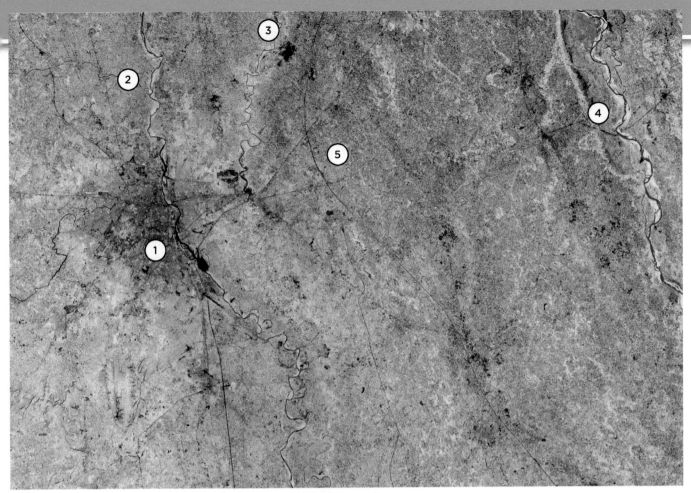

B Landsat 5 satellite image, 5 December 1989, New Delhi, India

C Landsat 8 satellite image, 5 June 2018, New Delhi, India

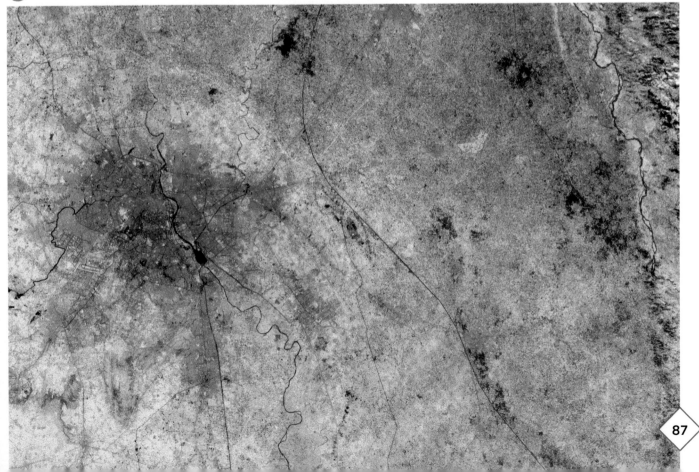

Learning objectives

▶ To compare and interpret news articles representing different viewpoints.

▶ To detect bias in news reports.

▶ To understand the significance of China's Belt and Road initiative.

▶ To use online newspapers to investigate world events and issues.

Newspapers and online news sites provide us with up-to-date information on events that take place locally and globally. Articles written by journalists are an important form of geographical data as they often include statistics, maps, graphs, opinions and eyewitness accounts. However, when using news articles as geographical data, we need to make sure we evaluate them carefully. Journalists usually approach a topic from a certain point of view and this can lead to bias. As a geographer, you can use your critical skills to identify and explain any bias, in order to fully evaluate sources of geographical data. In this lesson, you will begin to do this by analysing two articles published by news sites in the UK and China, on China's Belt and Road initiative.

What is China's *Belt and Road* initiative?

Beijing's multibillion dollar Belt and Road Initiative (BRI) is a state-backed campaign for global dominance and a massive marketing campaign for something that was already happening – Chinese investment around the world.

Over the five years since President Xi Jinping announced his grand plan to connect Asia, Africa and Europe, the initiative has morphed into a broad catchphrase. Belt and Road, or *yi dai yi lu*, is a '21st century silk road', confusingly made up of a 'belt' of overland corridors and a maritime 'road' of shipping lanes.

From South East Asia to Eastern Europe and Africa, Belt and Road includes 71 countries that account for half the world's population and a quarter of global GDP.

How much money is being spent?

The Belt and Road Initiative is expected to cost more than $1tn (£760bn), although there are differing estimates as to how much money has been spent to date […] But China's efforts abroad don't stop there. Belt and Road also means that Chinese firms are engaging in construction work across the globe on an unparalleled scale.

To date, Chinese companies have secured more than $340bn in construction contracts along the Belt and Road.

However, China's dominance in the construction sector comes at the expense of local contractors in partner countries […]

What are the risks for countries involved?

More recently, governments from Malaysia to Pakistan are starting to rethink the costs of these projects. Sri Lanka, where the government leased a port to a Chinese company for 99 years after struggling to make repayments, is a cautionary tale […]

Djibouti, Kyrgyzstan, Laos, the Maldives, Mongolia, Montenegro, Pakistan and Tajikistan are among the poorest in their respective regions and will owe more than half of all their foreign debt to China […]

Why is the initiative sparking global concern?

As Belt and Road expands in scope so do concerns it is a form of economic imperialism that gives China too much leverage over other countries, often those that are smaller and poorer.

Some worry expanded Chinese commercial presence around the world will eventually lead to expanded military presence. Last year, China established its first overseas military base in Djibouti. Analysts say almost all the ports and other transport infrastructure being built can be dual-use for commercial and military purposes.

'If it can carry goods, it can carry troops,' says Jonathan Hillman, director of the Reconnecting Asia project at CSIS.

A Article from *The Guardian* online, by Lily Kuo and Niko Kommenda, 30 July 2018, www.theguardian.com/cities/ng-interactive/2018/jul/30/what-china-belt-road-initiative-silk-road-explainer

China's integration with world will stay on course

It is an unarguable fact that the centre of world power is shifting toward the East at a quickened pace.

The share of developed economies, represented by the United States, Europe and Japan, in the global economy has dropped dramatically, as resource shortages, labour declines, shrinking markets, sluggish growth and other issues become more prominent. The US once accounted for half of the world's total economy, but the proportion fell to less than a quarter in 2018 …

In contrast, the developing countries, in particular China, are demonstrating bright prospects for economic growth, technological innovation and social development … The cooperation among the Asia-Pacific countries has seen sound progress … The world's economic and strategic centre is shifting toward the Asia-Pacific region …

Against this backdrop, the US has been targeting China. The country started and escalated the trade war with China … If the world's two largest economies are engaged in a strategic and historic race, the world may become divided. This might result in two isolated blocs coming into being.

In this scenario, unmanageable changes will occur in international relations, and overall stability of relations between countries will be hampered.

In this landscape, China, in particular, will play a vital role.

Through reform and opening-up, and promoting the Belt and Road Initiative, China has provided to the international community new philosophies such as development, security, cooperation, order, responsibility and civilisation.

China will also face changed international obligations and duties as well as mounting external pressure and challenges. However, China's integration with the world will stay on course and make progress in depth, scope and intensity.

B Article from *The China Daily Global* online, by Yu Hongjun, 6 August 2019,
www.chinadaily.com.cn/a/201908/06/WS5d48ce8aa310cf3e355640c5.html

Activities

1 a) How can news articles provide useful geographical data to investigate places?
   b) Why is it important to use news articles critically?
2 Read news articles A and B and copy and complete the following table to examine the purpose of each article.
3 Identify two examples of how each article seems to be biased. Justify your choices.
4 Write a paragraph that you think provides a balanced view of the Belt and Road scheme and its significance to world economic development.
5 Go to the Big Project online portal for world newspapers: www.thebigproject.co.uk/news/ You can use this site to track a major news story or world event as it happens, using different perspectives from newspapers in different countries.

	Who is the author?	Who published the article?	When was it published?	What facts are included?	What positive opinions are given?	What negative opinions are given?	What is the purpose of the article?
Article A							
Article B							

How can I use the internet to conduct a web enquiry about Mount Etna?

Learning objectives

▶ To describe the location of Mount Etna.

▶ To conduct web enquiries of Mount Etna.

Mount Etna is located at 37.73°N 15°E on the east side of Sicily. It is the highest and most active volcano in Europe, with a base circumference of 150 kilometres. It is a stratovolcano.

Two types of eruptions occur at Etna. Firstly, summit eruptions from one or more of the summit craters – historical lava flows from the craters, covering much of the surface of this massive volcano, many of which are marked, named and dated on map C. Secondly, eruptions at the sides of the volcano, which are less frequent, and can last a few hours or months, or even more than 20 years. These eruptions create cinder cones, hundreds of which are distributed on the slopes of Mount Etna (photo C).

Over one million people live within 30 km of the summit of Mount Etna, within the province of Catania. The fertile volcanic soils on the slopes of the volcano has attracted more than 20 per cent of Sicilians to live on its slopes. Major eruptions, happening at least once a decade, destroy homes and farm buildings; vines, olive groves and lemon trees vanish beneath lava and ash.

A Mount Etna from the central street of Catania

With thousands of people living in the vicinity of volcanoes, we wanted to better understand how they behave. Mount Etna has a long and frequent history of volcanic eruptions. It's a perfect natural laboratory, which we need to analyse thoroughly to protect the large population in the area.

B A volcanologist's view

C The hotel Rifugio Sapienza, near to the summit of Mount Etna

D The hotel from the cinder cone

Scientist use drones to investigate volcanoes

Scientists can now use drones to take thermal images and collect samples of gas from inside Italy's Mount Etna volcano. Previously scientists would climb the sides of the volcano on foot and wait for the wind to blow gas towards them for collection, which was obviously dangerous as the gas can be toxic, and the volcano could always erupt. Instead scientists can now fly relatively simple drones right into the volcano craters to better understand them, and develop plans for the evacuation of local residents should an eruption be expected.

Tourism and Mount Etna

The volcano was designated UNESCO World Heritage status in 2013 and is a popular destination for tourists, attracted to its unique landscape. Many people in the region earn a living as coach drivers, hauling tourists up the winding roads on the volcano's slopes, as well as at the restaurants and gift shops located close to the summit, as shown on photos C and D.

Activities

1 Write a paragraph to describe the absolute and relative location of Mount Etna. See Lesson 1.4 for guidance.

2 Write two sentences to describe the two ways Mount Etna erupts.

3 Look carefully at photo A and use the enquiry questions from Lesson 1.1 to write a paragraph to describe what the photo shows.

4 Go to Google Earth and search for Mount Etna.
   a) Compare the location of Catania and the volcano, and work out which direction the camera was pointing in photo A.
   b) Zoom in towards the volcano and find the hotel and cinder cone shown in photos C and D, work out in which direction the camera was pointing in photo C.
   c) Use the measuring distance tool (ruler) in Google Earth to find out how far the hotel is from the summit of the volcano. Why do you think the hotel is so close to the summit?
   d) Use Google street view to investigate the winding road, SP92, on the southern slope of the volcano. Identify five pieces of evidence to suggest that this road and the volcano are important for tourism.

5 Conduct a web enquiry beginning with an online video search for 'Mount Etna drone view'.
   a) Watch two video clips showing drone flights.
   b) Write three sentences explaining what you have discovered about the volcano from the drone flights.
   c) Search 'DJI Stories – predicting Mount Etna', and watch the video clip, which explains how scientists use drones to monitor the volcano.
   d) Using the videos and speech bubble B, identify three advantages of using drones to study volcanoes.

6 Conduct a web enquiry to identify why tourists are attracted to Mount Etna?
   a) Begin by conducting an online search using the enquiry question.
   b) Identify five reasons why tourist visit the volcano.

7 Write a paragraph to summarise what you have learnt about Mount Etna.

How can I investigate a volcanic eruption using data from remote sensing?

Learning objectives

▶ To interpret a range of data for a volcanic eruption.
▶ To interpret a satellite image of a volcanic eruption.

You first came across remote sensing in Lesson 10.2. Remote sensing is particularly useful for Earth scientists studying volcanoes, which tend to be both dangerous and inaccessible. In this lesson, you will investigate a recent eruption at Mount Etna by interpreting evidence from remote sensing.

This volatile volcano is monitored by the Etna Observatory, part of Italy's National Institute of Geophysics and Volcanology. On 24 December 2018, a swarm of 130 earthquakes occurred over a three-hour period on the side of Mount Etna (see map C). An eruptive fissure opened at the eastern base of the South-East Crater, generating a cloud of ash, visible several kilometres away – see photos A, B and D. Ash covered local villages and delayed aircraft from landing at the nearby Catania airport. Lava flowed from the fissure between 26 and 27 December (B). According to Boris Behncke, a volcanologist at the Etna Observatory, this was the first side eruption at Etna for more than ten years. He recorded images of the event on his Twitter account, D.

A Mount Etna eruption, 27 December 2018, 12.49, VIIRS satellite

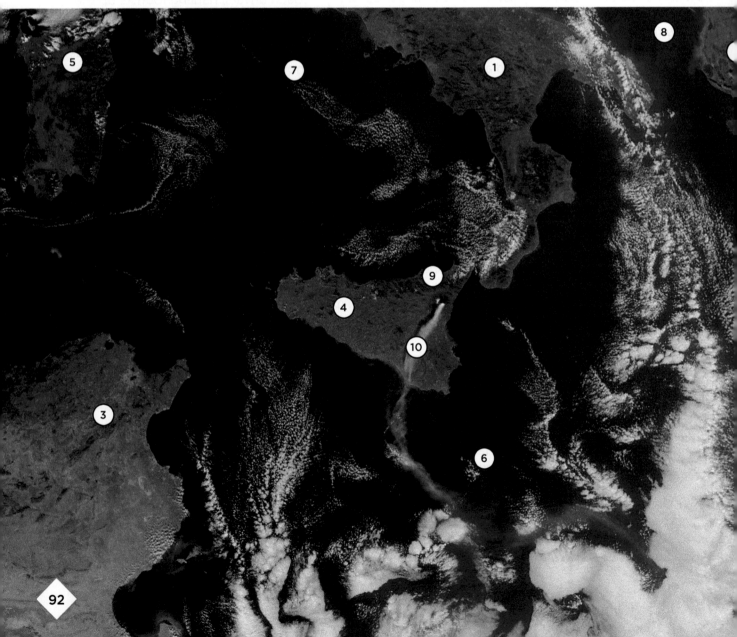

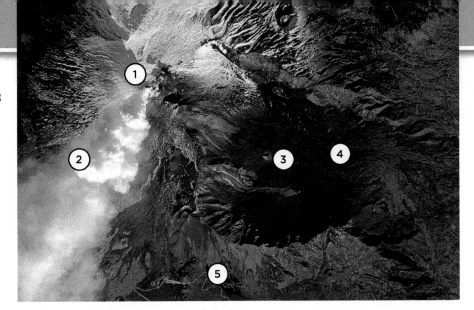

B Mount Etna summit,
28 December 2018, Landsat 8

C Epicentre map of earthquakes recorded on Mount Etna on 24 December 2018

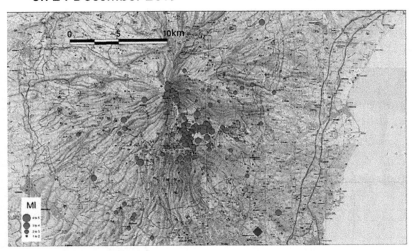

Boris Behncke @etnaboris 24 December 2018

Start of #Etna's first flank eruption for more than 10 years, shortly after noon on 24 December 2018, seen from my home in the village of Tremestieri Etneo, 20 km south of the summit of the volcano.

D Boris Behncke's tweet,
24 December 2018

Activities

1 Why is remote sensing important in studying volcanoes?

2 When did the side eruption of Mount Etna occur?

3 Look carefully at satellite image A and compare it with an atlas map of Europe.
   a) Name the: countries 1–3; islands 4–5; seas 6–8; landform 9; feature 10.
   b) In which direction was the wind blowing when the image was taken?
   c) What evidence suggests the wind direction changed over the sea?

4 Look carefully at satellite image B and think back to your research using Google Earth from Lesson 11.1
   a) Write a paragraph to describe what is happening at locations 1–3.
   b) Name the landforms at location 4.
   c) Name the human features at location 5.

5 Look carefully at map C.
   a) Write a paragraph to describe the distribution of earthquakes.
   b) Identify a reason why these earthquakes may have occurred.

6 Look carefully at tweet D.
   a) Who is Boris Behncke?
   b) Why does he think this eruption is so significant?
   c) Describe what his photo shows, and where it was taken.

7 Identify three things you have learnt about the eruption from the data you have examined in this lesson.

Learning objectives

▶ To understand what an earthquake is.

▶ To understand where earthquakes occur.

▶ To be able to use a GIS to investigate earthquakes.

An earthquake is what happens when two blocks of the Earth suddenly slip past one another. The surface where they slip is called the fault or fault plane. The location below the Earth's surface where the earthquake starts is called the focus, and the location directly above it on the surface of the Earth is called the epicentre (see diagram B).

Earthquakes are recorded by instruments called seismographs. The recording they make is called a seismogram. The size of the earthquake is called its magnitude. Earthquakes are recorded by a seismographic network. Each seismic station in the network measures the movement of the ground at that site. The slip of one block of rock over another in an earthquake releases energy that makes the ground vibrate. That vibration pushes the adjoining piece of ground and causes it to vibrate, and thus the energy travels out from the earthquake epicentre in a wave. The United States Geological Survey (USGS) maintains an Earthquake Hazards Programme, to better understand the occurrence of earthquakes and their effects and consequences. Screenshot A shows the real-time GIS interactive map that plots earthquakes around the world as they occur.

Ⓐ **USGS GIS Earthquake map tool guide,** https://earthquake.usgs.gov/earthquakes/map

The latest earthquake to occur around the world is updated here and added to the map.

Click on plus and minus tool to zoom in and out of the map.

Click on the 'Settings' tool to change the layers and settings on the map, including the format, order and layers.

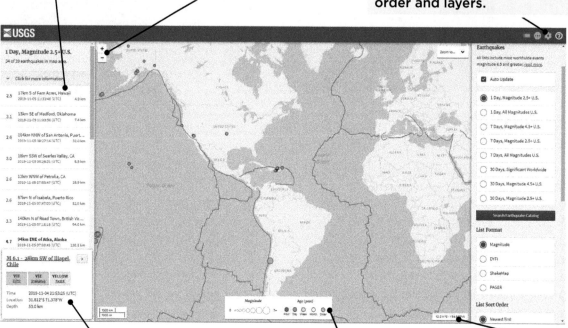

When you click on an earthquake on the map this box appears to provide a link to another webpage that provides more information about this earthquake. This includes another interactive GIS map and regional information on the tectonic history of the region.

Click on the show legend box to see this legend or key. The size of circle on the map shows the magnitude of the eartquake, the colour of the circle indicates when it occured.

As you move the cursor across the map, the latitude and longitude are shown here.

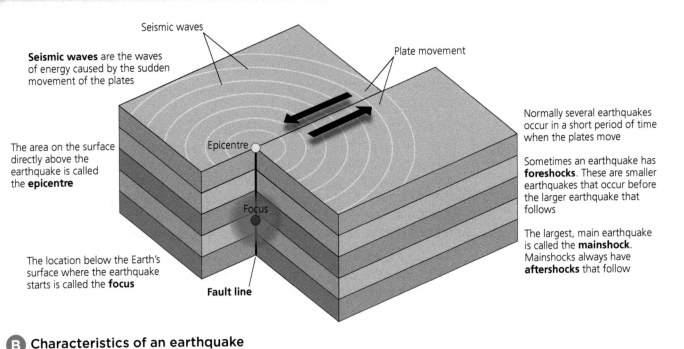

Seismic waves

Seismic waves are the waves of energy caused by the sudden movement of the plates

Plate movement

The area on the surface directly above the earthquake is called the **epicentre**

Epicentre

Focus

The location below the Earth's surface where the earthquake starts is called the **focus**

Fault line

Normally several earthquakes occur in a short period of time when the plates move

Sometimes an earthquake has **foreshocks**. These are smaller earthquakes that occur before the larger earthquake that follows

The largest, main earthquake is called the **mainshock**. Mainshocks always have **aftershocks** that follow

B **Characteristics of an earthquake**

Activities

1 Look carefully at diagram B.
   a) What is an earthquake?
   b) Draw your own diagram to show the main features of an earthquake and how they occur. Annotate an explanation of the following on your sketch: fault, epicentre, focus, foreshock, main shock, aftershock, seismic waves.
   c) How do scientists measure earthquakes and record their size?

2 a) What is the USGS?
   b) What is the purpose of their Earthquake Hazard Programme?

3 Go to the USGS's online earthquake GIS map: https://earthquake.usgs.gov/earthquakes/map
   a) Identify how many earthquakes have occurred today. Write a list of the magnitude, place and location of the ten strongest quakes.
   b) Use the 'Settings' tool and change the map layer to 'Satellite'.
   c) Use the minus tool and zoom out to a global view.

   d) Write a paragraph to describe the distribution of earthquakes for the day.
   e) Go to the 'Settings' tool and change the earthquake data to '7 Days, All Magnitudes'. Add to your paragraph two new sentences to describe any changes in the distribution of earthquakes.
   f) Go to the 'Settings' tool again. Scroll down to 'Map Layers' and click the box to add 'Plate boundaries'. Now compare the distribution of earthquakes and add three sentences to your paragraph to explain the relationship between the distribution of earthquakes and the boundaries.

4 Select one of the earthquakes from the list to the left of the map – click on a place that had an earthquake of a magnitude 4 or above. Click on the box that appears. Use the information about the history of earthquakes in this area, and maps showing plate boundaries, to explain why this area suffers from earthquakes.

How can I use atlas maps and climate data to investigate the climate of Africa?

Learning objectives

▶ To be able to draw and compare climate graphs.

▶ To interpret isotherm and isohyet atlas maps of Africa.

▶ To describe the climate of Africa.

The climate of Africa is more variable by precipitation than by temperatures, which are mostly high. The continent mainly lies between the tropics, leading to the dominance of warm and hot climates.

Maps A, B and D are atlas maps that show the climate of Africa using lines on a map that join places of equal value, similar to contour lines on topographical maps. Maps A and B show isotherms, which join together places of the same temperature, and Map D shows isohyets, which join together places of the same precipitation. These maps also use colour to show areas of the same temperature and rainfall. These lines and colours clearly show patterns of climate.

An atlas usually includes a climate page showing these types of maps for each continent, often with climate data (see C), or graphs for different places. In this lesson you will investigate the climate of Africa using the data and maps.

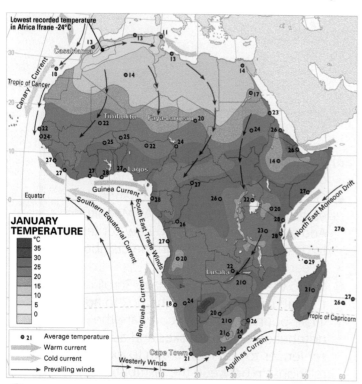

A January temperatures in Africa, isotherm map, © *Philip's Modern School Atlas*

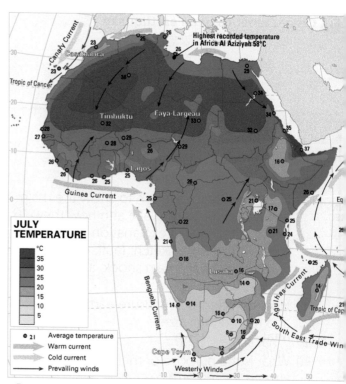

B July temperatures in Africa, isotherm map, © *Philip's Modern School Atlas*

			J	F	M	A	M	J	J	A	S	O	N	D
Casablanca, Morocco		**Temp, °C**	13	13	15	16	18	21	23	23	22	20	17	14
59 m	33.57°N 7.58°W	**Rain, mm**	78	61	54	37	20	3	0	1	6	28	58	94
Timbuktu, Mali		**Temp, °C**	22	25	28	31	34	34	32	30	31	31	28	23
269 m	16.76°N 3.00°W	**Rain, mm**	0	0	0	1	4	20	54	93	31	3	0	0
Faya-Largeau, Chad		**Temp, °C**	20	22	26	30	33	34	33	33	33	29	24	21
245 m	17.92°N 19.11°E	**Rain, mm**	0	0	0	0.1	0.4	0.3	3	7	0.8	0.1	0	0
Lagos, Nigeria		**Temp, °C**	27	28	28	28	27	26	25	24	25	26	27	27
40 m	6.52°N 3.37°E	**Rain, mm**	28	41	99	99	203	300	180	56	180	190	63	25
Lusaka, Zambia		**Temp, °C**	22	22	21	21	18	17	16	19	22	25	23	22
1,154 m	15.38°S 28.32°E	**Rain, mm**	224	173	90	19	3	1	0	1	1	17	85	196
Cape Town, South Africa		**Temp, °C**	21	20	20	17	14	13	12	12	14	16	18	20
44 m	33.92°S 18.42°E	**Rain, mm**	12	19	17	42	67	98	68	76	36	45	12	13

C Climate data for six African cities

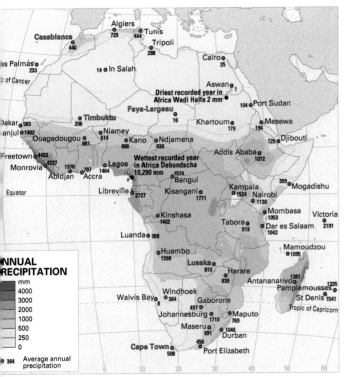

D Annual precipitation in Africa, isohyet map,
© *Philip's Modern School Atlas*

Activities

1 Write definitions for isotherms and isohyets.
2 Work with a partner and refer back to Lesson 4.3 if you need help.
   a) Each of you choose three cities from A and draw three climate graphs to represent the data given.
   b) For each of your cities, calculate the following and label it on your graphs: average monthly temperature; temperature range; annual rainfall.
   c) Write a paragraph for each graph describing the climate for each place.
   d) Compare your findings with your partner for the other three locations.
3 Draw a table with the following rows to summarise key climate data for each of the cities in C:
   Max. monthly temp.; Min. monthly temp.; Average monthly temp.; Annual rainfall; Month with highest rainfall; Month with lowest rainfall.
4 Write a paragraph underneath your table, comparing the climate of cities.
5 Identify the locations of the climate graphs on maps A, B and D.
   a) Describe the distribution of temperature on maps A and B, climate graphs and your table.
   b) Describe the distribution of rainfall on map D, climate graphs and your table.
6 Write a concluding paragraph to summarise what you have discovered about climate in Africa, using the maps and graphs.

Enquiry skills: How can I investigate how the Lake Chad region of Africa is changing? Part 1

Learning objectives

▶ To ask geographical questions.

▶ To analyse a range of geographical data, using enquiry questions.

▶ To identify causes and consequences of change.

▶ To consider decisions about future solutions to problems.

Lake Chad was one of the largest fresh waterbodies in Africa, about 8% of the size of Africa. In the last 50 years, 90 per cent of Lake Chad has disappeared. This has caused great changes in the whole region as shown in satellite images B and C.

In this lesson, you will consider a range of geographical data about the problems, their causes and a possible scheme to deal with these issues. You will use your geographical enquiry skills to summarise these problems and evaluate the suggested schemes.

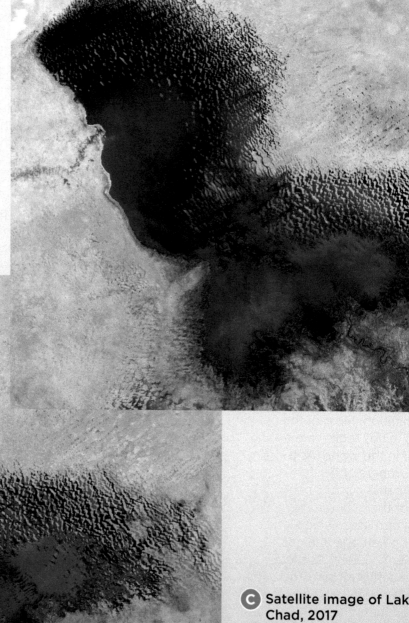

B Satellite image of Lake Chad, 1973

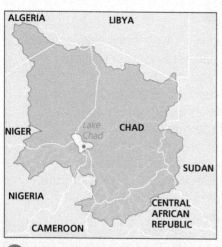

A Map of Lake Chad Basin

C Satellite image of Lake Chad, 2017

Due to climate change and high demand for water, Lake Chad has shrunk to a twentieth of the size it was in 1963. The result is degraded ecosystems, water shortages, crop failures, livestock deaths, collapsed fisheries, increased soil salinity and, as a result, increased poverty. Extremist groups, most notably Boko Haram, have developed in the region. They have claimed more than 35,000 lives and caused the displacement of 1.8 million people from north-eastern Nigeria.

D United Nations Development Programme

Desertification really scares us. We see more and more sand blowing into the area every year. The air is dusty, the wind is fierce and unrelenting, the plants are wilting and the earth is turning into sand dunes. The sparse vegetation is occasionally broken by withered trees and shrubs. It's increasingly difficult to grow crops. The lake area is dying.

E Farmer

With water levels at their lowest, combined with a larger-than-ever population pulling from the ground water for personal use or for irrigation projects, the lake is not able to replenish itself.

F Aid worker

The lives of all of us herders, fisherfolk and farmers are threatened as the lake dries up before our eyes. Many have moved towards greener areas, to try to farm. Others have gone to Kano, Abuja, Lagos and other big cities to try a different life.

G Village elder

The Chad Basin lies within the Sahel, a semiarid strip of land dividing the Sahara Desert from the humid savannas of equatorial Africa. A number of countries share the basin, see map A. An estimated population of 40 million depend on the lake for crop and livestock farming, fishing, commerce and trade. The lake is also a source of water supply for drinking, and sanitation. Most of the countries located in this region are among the poorest in the world. As the lake shrank, large numbers of herders had to search for alternative pastures and sources of water for their cattle. They started to graze existing farmland, which led to disputes over crop damage and cattle theft that mostly turn violent. As the herders are predominantly Muslim and the farmers largely Christian, religious radicals have exploited the conflict, see H.

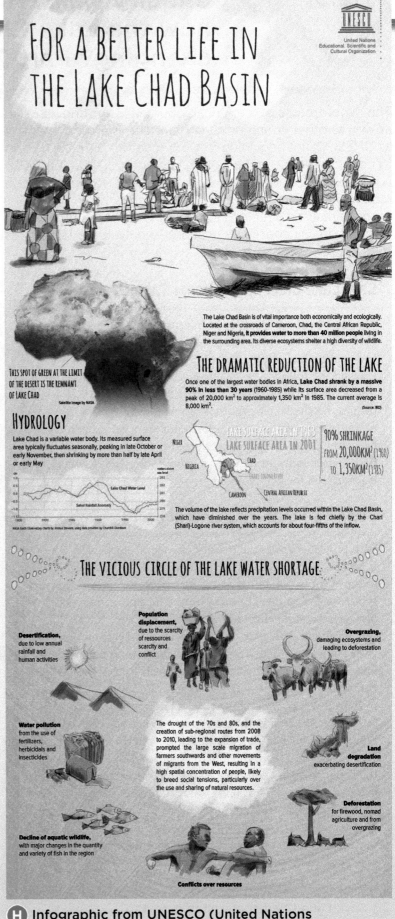

FOR A BETTER LIFE IN THE LAKE CHAD BASIN

UNESCO
United Nations
Educational, Scientific and
Cultural Organization.

The Lake Chad Basin is of vital importance both economically and ecologically. Located at the crossroads of Cameroon, Chad, the Central African Republic, Niger and Nigeria, **it provides water to more than 40 million people** living in the surrounding area. Its diverse ecosystems shelter a high diversity of wildlife.

THIS SPOT OF GREEN AT THE LIMIT OF THE DESERT IS THE REMNANT OF LAKE CHAD
Satellite image by NASA

THE DRAMATIC REDUCTION OF THE LAKE

Once one of the largest water bodies in Africa, **Lake Chad shrank by a massive 90% in less than 30 years** (1960-1985) while its surface area decreased from a peak of 20,000 km² to approximately 1,350 km² in 1985. The current average is 8,000 km².

(Source: IRD)

HYDROLOGY

Lake Chad is a variable water body. Its measured surface area typically fluctuates seasonally, peaking in late October or early November, then shrinking by more than half by late April or early May

Lake Chad Water Level
Sahel Rainfall Anomaly

NASA Earth Observatory charts by Joshua Stevens, using data provided by Churnill Okonkwo

LAKE SURFACE AREA IN 1963
LAKE SURFACE AREA IN 2001

NIGER
NIGERIA
CHAD
CHARI-LOGONE RIVER
CAMEROON
CENTRAL AFRICAN REPUBLIC

90% SHRINKAGE
FROM 20,000KM² (1960)
TO 1,350KM² (1985)

The volume of the lake reflects precipitation levels occurred within the Lake Chad Basin, which have diminished over the years. The lake is fed chiefly by the Chari (Shari)-Logone river system, which accounts for about four-fifths of the inflow.

THE VICIOUS CIRCLE OF THE LAKE WATER SHORTAGE

Desertification, due to low annual rainfall and human activities

Population displacement, due to the scarcity of ressources scarcity and conflict

Overgrazing, damaging ecosystems and leading to deforestation

Water pollution from the use of fertilizers, herbicidals and insecticides

The drought of the 70s and 80s, and the creation of sub-regional routes from 2008 to 2010, leading to the expansion of trade, prompted the large scale migration of farmers southwards and other movements of migrants from the West, resulting in a high spatial concentration of people, likely to breed social tensions, particularly over the use and sharing of natural resources.

Land degradation exacerbating desertification

Deforestation for firewood, nomad agriculture and from overgrazing

Decline of aquatic wildlife, with major changes in the quantity and variety of fish in the region

Conflicts over resources

H Infographic from UNESCO (United Nations Educational, Scientific and Cultural Organisation)

The governments of the African countries surrounding Lake Chad, as well as other countries and organisations around the world, such as UNESCO, are beginning to work together to make decisions to help the people in the Lake Chad region. The solutions to this complex crisis of environmental, social and economic problems require leaders to consider evidence from a wide variety of experts to make realistic decisions to begin to overcome problems. In this lesson, you will consider a proposal and evaluate its potential.

Save Lake Chad International Conference report

On 26–28 February 2018, an International Conference on Lake Chad was held in Abuja, Nigeria to discuss options and proposals about the future of the region. The delegates reached the following conclusion: 'There is no solution to the shrinking of Lake Chad that does not involve recharging the lake by transfer of water from outside the basin. Inter-basin water transfer is not an option, but a necessity.' This conference endorsed the Transaqua Project, first suggested by an Italian company in the 1980s, when it was rejected for being too large.

Overview of the Transaqua project

This project involves the transfer of water from the Congo River basin to Lake Chad, by building a 2400 km canal to move about 100 billion cubic metres of water from the River Congo to the lake every year. This only represents between 4 and 8 per cent of water of the Congo River. The water would be diverted downslope by gravity, from the river via a series of 20 steps, where a dam would be built. At many of these dams, hydro-electric power stations would be constructed, with the potential to generate 15,000–25,000 million KWh of hydroelectricity and irrigate 50,000 to 70,000 km² of land in the Sahel zone. The project would require the collaboration of 12 African countries.

A delegation from the Italian company Bonifica made the proposal at the conference and has offered to donate €1.6m to conduct a feasibility study. Chinese company PowerChina has offered to match this funding. The two companies have since signed an agreement to cooperate on the scheme. It is currently estimated that the scheme would cost between $50 billion and $70 billion.

I Save Lake Chad conference report

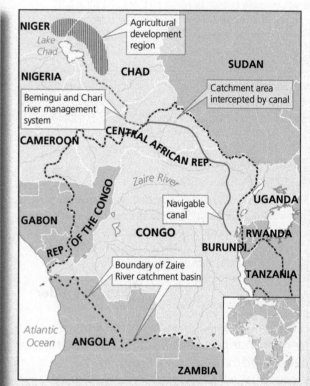

J Map of Transaqua Project

K Farmer

Two major problems are the cost of the scheme – where will this amount of money come from? But most importantly, the proposal predicts that it will take 30 years (until 2047) before the lake has a noticeable increase in water. That is much too slow to stop Lake Chad drying up.

The canal will be a major geographical barrier to endangered wildlife in the region.

L An environmentalist

Fanciful, it costs too much, but does it really?! A recent economic study found that in the last 15 years, in 23 African countries, $284 billion has been spent in conflicts. That figure of $284 billion only counts structures that have been destroyed, health-care costs and costs linked to refugees. Then there are the other costs that are not counted: managing the refugees, decline of trade and political instability, so more like $300 billion – that's $20 billion per year, from 1990 to 2005, on negative outcomes. This puts the $50–70 billion for our proposal in a very different light.

We, the Democratic Republic of the Congo (DRC), are opposed to the Transaqua project. The DRC alone is the source of the largest part of the waters of the Congo Basin, to which the Ubangi River belongs, and thus its approval is required before any go-ahead to the transfer project. We must protect our own use of this water.

M A DRC government minister

N An Italian businessman

Activities

1 Look carefully at the range of geographical data, A–H, in Lesson 12.2.
  Write three sentences to answer each of the enquiry questions about Lake Chad. Quote evidence from the geographical data to support your answers in each case.
  – **Where** is this place?
  – **What** is it like?
  – **Why** is it like this?
  – **How** is it changing?
  – **Who** is affected by the changes?
  – How do you **feel** about this place?

2 Look carefully at satellite images B and C. Write a paragraph to describe how they show that Lake Chad is changing.

3 a) Identify which geographical data provides evidence of what is causing Lake Chad to change.
  b) Write a paragraph to explain these causes.

4 a) Identify which geographical data provides evidence of the consequences of the changes to Lake Chad.

  b) Write a paragraph to explain these consequences.

5 Study map J and read conference report I.
  a) Write four sentences to describe the key elements of the Transaqua Project.
  b) Look back at the climate maps for Africa in Lesson 12.1. Write two sentences to describe the climatic evidence to support the project.
  c) Identify three advantages of the project for the Lake Chad region.

6 Read the different viewpoints, K–N, about the scheme.
  a) Categorise them into positive and negative views of the scheme.
  b) List the disadvantages of the scheme.

7 a) Work in a group of four and discuss the advantages and disadvantages of the scheme for the future of the Lake Chad region.
  b) Make a decision about whether the idea offers the best future for the region.
  c) Write a paragraph to justify your choice.

Learning objectives

▶ To locate photographs on OS maps of different scales.

▶ To compare ground-level photos with an OS map.

▶ To identify geographical features on OS maps at different scales and photos.

Cwm Idwal is located at 53.12° N 4.03° W, in the Snowdonia National Park, in North Wales. The area is popular with visitors walking along a footpath trail, maintained and promoted by the National Trust and the National Park. A family planned a walk in the area, using OS map A, 1:10 000, and map-flap A, 1:25 000, to follow a route. The route is marked on A, with numbers 1–8, where the family stopped to take photographs B–E and G–J. Other families shared their thoughts about the Cwm Idwal walk, with reviews on a tourist website, F and K.

B Footbridge (FB) over river from Llyn Idwal (llyn is Welsh for 'lake')

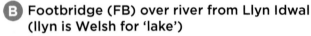

A Cwm Idwal walk, 1:10 000 OS map

© Crown copyright and database rights 2020 Hodder Education under licence to Ordnance Survey

E Footpath looking towards Cwm Idwal

C Ogwen Cottage

D Milestone (.MS) on the A5

Spectacular views on well-made path (total ascent is around 130 metres and is all very gradual). The walk takes around 2hrs at a slow pace. There is a small National Trust car park with toilets and a cafe/snack bar sign-posted from the main road (A5) at the western end of Llyn Ogwen. However, when we visited on Good Friday there were hundreds of cars parked on either side of the A5.

F Visitor review

Activities

1 Use maps on your phone or computer to plan a route from your house to Cwm Idwal. How many miles is it and how long would the journey take by car? Describe the route suggested by the map software.

2 Look carefully at map A, map-flap A and photos B–E and G–J.
   a) Match the photos to the stops on the route.
   b) Give the six-figure grid reference for each location.
   c) In each case, provide map evidence to justify your choice.

3 Rewrite the description of the route followed, adding the six-figure grid reference, the direction they walked at each point and descriptions from photos B–E and G–J.

4 Use map-flap A and the linear scale to measure the distance the family walked. Add this to your description of the route.

5 Draw a fieldsketch of Cwm Idwal, photo I.
   a) Label the following on the sketch: Llyn Idwal; the steep backwall; Devil's Kitchen; the footpath; the smoothed rock in the left-hand corner of the photo.
   b) Add a title that includes the direction the camera was pointing and the six-figure grid reference for where the photo was taken.

6 Read the reviews of visitors, F and K.
   a) What problem do visitors often experience when arriving by car?
   b) What health and safety factors need to be considered before walking the route?

The family wanted to park their car at Ogwen Cottage, but when they arrived the car park was full, so they eventually parked at the first car park on the A5 on the southern side of Llyn Ogwen. The family walked along the A5 and passed Ogwen Cottage, to get a view of Ogwen Waterfall. On the way they noticed a milestone at the side of the road. They then visited the toilets before starting the trail. They crossed the footbridge and admired the smaller waterfall. Halfway along the footpath, they took a photo looking up to Cwm Idwal. They had a drink of water and admired the view at Llyn Idwal, before returning along the path and A5 back to their car.

J Ogwen Waterfall from the A5

G A5 alongside Llyn Ogwen, halfway from first car park to Ogwen Cottage

H Entrance to Ogwen car park

I Cwm Idwal and Llyn Idwal

Went up here for a geography course many years ago and decided to come here again. It is simply breathtaking. The area is beautiful. The paths are uneven and can be slippery when wet. I think that walking boots are a must and do check the weather as cloud can descend quickly resulting in poor visibility. It's not a place to visit wearing flip-flops!

K Visitor review

Learning objectives

▶ To compare a vertical aerial photo with an OS map of the same scale.

▶ To identify features and land uses on the map and aerial photo.

▶ To draw a sketch map to record evidence of glacial erosion features.

Photo A is a vertical photograph of Cwm Idwal and the Nant Ffrancon valley in North Wales, showing the same area as map-flap A. Both the photo and the OS map are at the same scale. You can use maps and photos like these to identify both physical and human geographical features.

During the last ice age, 10,000 years ago, much of Snowdonia was covered in ice. On the colder north and east-facing slopes, sheltered from the Sun, ice built up in hollows. Ice formed in these hollows, eventually rotating out, over-deepening the hollows to form corries, which are called cwms in Wales. A succession of cwms formed along the western slopes of the Nant Ffrancon area. These corrie glaciers merged together downslope, following the existing valley, forming a glacier, which created the U-shaped valley of Nant Ffrancon. There are large numbers of corries, called cwms on map-flap A; you can also clearly see their shapes on aerial photo A.

Activities

1 Look carefully at photo A and map-flap A.
   a) Locate the ten features labelled A–J on photo A.
   b) Copy and complete the table to record each feature A–J.

Feature	Name of feature	Six-figure grid reference
Lake A		
Lake B		
Lake C		
Lake D		
River E		
Valley F		
Land use G		
Land use H		
Land use I		
Road J		

2 Draw a sketch map to show the area shown in photo A and map-flap A.
   a) Begin by drawing and numbering all the grid squares.
   b) Draw the Nant Ffrancon valley floor.
   c) Draw the shapes of the seven corries, or cwms, to the west of the valley, and Cwm Idwal to the south.
   d) Name the cwms on your sketch.
   e) Add and label Llyn Ogwen on your sketch. You will get the chance to add more features of glacial erosion to your sketch map in Lesson 13.3.

3 Go back to your fieldsketch of Cwm Idwal from Lesson 13.1.
   a) Label the characteristic features of a corrie on your sketch: steep backwall; hollow filled by lake; Llyn Idwal ('llyn' is Welsh for lake); lip.
   b) Annotate your sketch to explain how the corrie was formed.

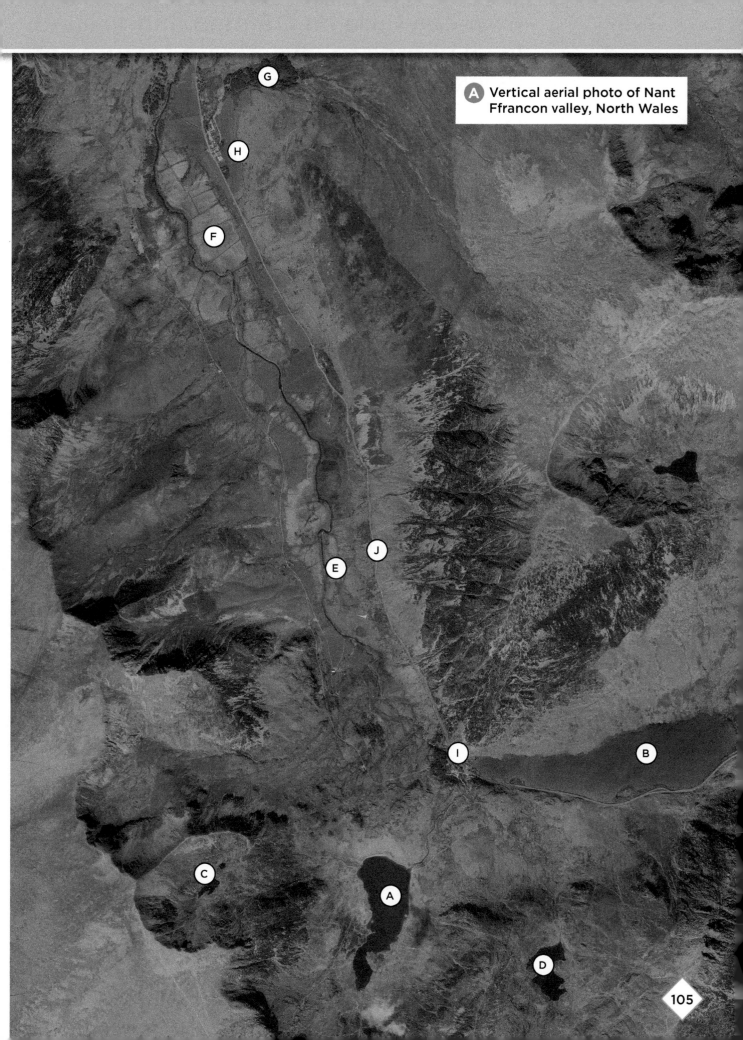

Vertical aerial photo of Nant Ffrancon valley, North Wales

How can I use an OS map and ground-level photos to investigate features of glacial erosion?

Learning objectives

▶ To compare ground-level photos with an OS map.

▶ To identify glacial features on a map and on photos to explain the impact of a valley glacier.

▶ To draw a cross-section to show a glacial feature.

Geographers have attempted to reconstruct the path of glaciers by studying the features and landscape they have left behind. Physical geographers have investigated all the evidence of glaciation in the Nant Ffrancon valley to construct model A, to show how they think ice flowed down the valley to form the landforms. Nant Ffrancon was created by a valley glacier. It is a U-shaped valley, C, which has steep sides and a flat valley floor. Today, a misfit stream, far too small to have created the valley, meanders its way across the valley floor. The sides of the valley are marked by truncated spurs, C, which are the remnants of the interlocking spurs of the original river valley. These were cut off as the ice created a straighter path.

Along the sides of Nant Ffrancon there are also hanging valleys, B. As shown in A, these were formed as a result of the main valley being over-deepened by the main ice, creating the steep slopes of the valley sides. Today, tributary streams fall from the slopes of the valley sides. Llyn Ogwen is a ribbon lake, which fills a glacial valley floor. It extends from the lip of the Ogwen Falls in the glacial hollow.

In this lesson, you will use map-flap A and ground-level photos, B–D, to identify some of these features to consider the path of the glaciers.

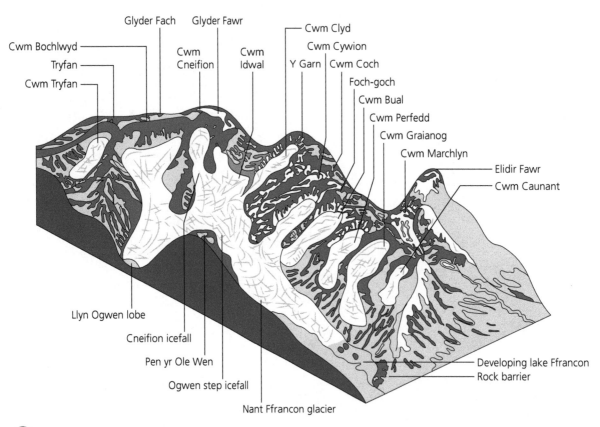

A A model of Nant Ffrancon valley during the last ice age

B Hanging valley taken from the A5 at 644618

C Nant Ffrancon valley taken at 647603

D Nant Ffrancon valley taken from the A5 at 644623

Activities

1 Look carefully at diagram A.
   a) How were geographers able to visualise what the Nant Ffrancon valley looked like during the last ice age?
   b) Write three sentences to describe how glaciers flowed through the area.

2 Compare photo B with map-flap A.
   a) In which direction was the camera pointing when the photo was taken?
   b) Write three sentences to describe how the feature shown in the photo was formed.

3 Compare photo C with map-flap A.
   a) In which direction was the camera pointing when the photo was taken?
   b) Identify features 1–3 labelled on the photo.
   c) Draw a fieldsketch of this view. Label your answers to a) and b) on your sketch.

   d) Annotate your fieldsketch to explain how the valley was formed.

4 Draw a cross-section across Nant Ffrancon from Y Galan at 632622 to Braich Ty Du at 653624. Label the features 1–3 shown on photo C on your cross-section.

5 Compare photo D with map-flap A.
   a) In which direction was the camera pointing when the photo was taken?
   b) Identify features 4–6 labelled on the photo.

6 Go back to your sketch map from Lesson 13.2. Add the features investigated in this lesson to your map.

7 Write a paragraph to review what you have discovered about glacial features in the area in Lesson 13.2 and this lesson. Explain how the evidence helps you to understand how glaciers created the landforms.

Enquiry skills: How can I test a hypothesis about glacial erosion using an OS map and a GIS?

Learning objectives

▶ To test a hypothesis as part of geographical enquiry.

▶ To present data using a rose diagram.

▶ To analyse geographical patterns on an OS map.

▶ To use an online GIS to test a hypothesis about corries.

Hypothesis testing is an important element of geographical enquiry. A **hypothesis** is an idea or explanation that can be tested through investigation. This is often best done through fieldwork, but you can also use OS map evidence.

In Lesson 13.2, you produced a sketch map to show the position of corries, or cwms, in the Nant Ffrancon valley. Geographers generally believe that aspect, the compass direction that a slope faces, influenced the development of corrie glaciers on the north and east flanks of mountains in the UK. These slopes faced away from the Sun, thereby allowing snow and ice to collect on the slope sheltered from the Sun's heat, forming in hollows. Eventually the weight of the built-up ice led to it rotating out of the now enlarged and over-deepened hollow, leaving behind the characteristic features of a corrie: steep backwall, hollow and lip. West and south-facing slopes tend to be warmer, so ice did not build up in the same way.

In this lesson, you will initially test this hypothesis using your sketch map and map-flap A, to draw a **rose diagram**, diagram A, before widening your investigation using an online GIS of glacial features. A rose diagram is a circular histogram used to plot directional data such as wind direction, when investigating weather, or in this case, the direction corrie glaciers are facing.

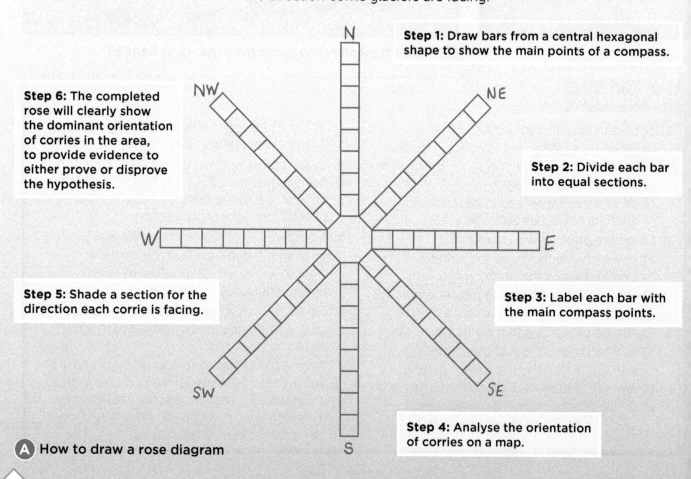

Step 1: Draw bars from a central hexagonal shape to show the main points of a compass.

Step 2: Divide each bar into equal sections.

Step 3: Label each bar with the main compass points.

Step 4: Analyse the orientation of corries on a map.

Step 5: Shade a section for the direction each corrie is facing.

Step 6: The completed rose will clearly show the dominant orientation of corries in the area, to provide evidence to either prove or disprove the hypothesis.

A How to draw a rose diagram

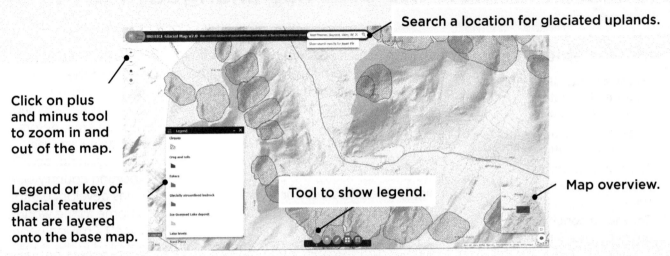

Click on plus and minus tool to zoom in and out of the map.

Search a location for glaciated uplands.

Legend or key of glacial features that are layered onto the base map.

Tool to show legend.

Map overview.

B University of Sheffield BRITICE GIS, www.BRITICEmap.org

The University of Sheffield and BRITICE

The University of Sheffield has developed a GIS called BRITICE, with Esri UK, using ArcGiS (see screenshot B). This online GIS database plots over 170,000 landforms from over 100 years of field investigation, for the glaciated landscape of the British Isles. This provides an excellent picture of the distribution and pattern of Britain's glacial landforms. By investigating the pattern of the retreat of glaciers in the UK, it is possible to date the timing and speed of ice retreat during a former climate change. So far you have only tested the hypothesis for one area for a limited number of corries. You can use this online GIS to further your hypothesis testing, to obtain more meaningful results.

Activities

1 What is hypothesis testing?
2 What do geographers mean by the word 'aspect'?
3 In your own words, write out the hypothesis to be tested in this lesson.
4 Look carefully at map-flap A and your sketch map from Lesson 13.2.
   a) Count and record the number of corries that surround and formed the Nant Ffrancon valley.
   b) Decide which direction each corrie is facing or orientated.
   c) Draw a rose diagram, using the guidance provided in A and record the direction of corries
   d) Add any other corries you can see on map-flap A.
   e) Write a paragraph to describe your results.
   f) Write a conclusion to identify whether the hypothesis is proved or not.
5 What is the BRITICE GIS?

6 Why has the University of Sheffield developed this GIS?
7 Go to the online GIS, using screenshot B to help you navigate the data.
   a) Start by searching the Nant Ffrancon valley. Use the legend to identify how glacial features in particular corries are shown.
   b) Search another glaciated area of the UK – start with Helvellyn in Cumbria.
   c) Repeat Question 4.
   d) If you have time, repeat the process for a glaciated area in Scotland.
8 Write a paragraph to explain the advantage of using the GIS to compare the orientation of corries in different parts of the UK.
9 Conclusion: You should now have two or three rose diagrams.
   a) Compare the patterns they show.
   b) Write a paragraph to explain whether the hypothesis is proved or disproved.

Enquiry skills: How can I investigate how the UAE has changed? Part 1

Learning objectives

▶ To use a range of geographical data to conduct an enquiry.

▶ To be able to draw a graph to show population change.

▶ To be able to draw a population pyramid for the UAE.

▶ To reach a conclusion to answer an enquiry question.

In the three decades since 1965, the United Arab Emirates (UAE) has developed from little more than a desert into one of the most developed countries in the Middle East. The country was formed in 1971, after the discovery of vast oil reserves. It is a group, or federation, of seven 'emirates' – land ruled by a monarch called an emir. Before the discovery of oil, much of the economy of the Middle East was dominated by nomadic farming, date palm cultivation, fishing, pearling and seafaring. Today, the UAE is a modern, oil-exporting country with a highly diversified economy. Dubai, in particular, has developed into a global hub for tourism, retail and finance and is home to the world's tallest building and largest man-made seaport. Over 9.7 million people live in the country, with nearly 85 per cent concentrated in the three largest emirates – Abu Dhabi, Dubai and Sharjah. In this lesson, you will use and research a wide range of geographical data to investigate what the country is like and how it has changed.

A The UAE in the 1970s

B Dubai – a world city

C Atlas map of the UAE

Conduct your own internet research

You can use the geographical data provided in Lessons 14.1 and 14.2. You can also conduct your own research. You will find the following websites very useful:

● http://confluence.org/

● Google Earth

● the Dubai tourist website: www.visitdubai.com/en

● ArcGIS storymap – a two-minute time lapse of Dubai, presents a wide range of geographical data, including video clips satellite images, photos and maps, to show how Dubai has changed. Find it using a search engine typing in 'ArcGiS storymap two-minute time lapse of Dubai'.

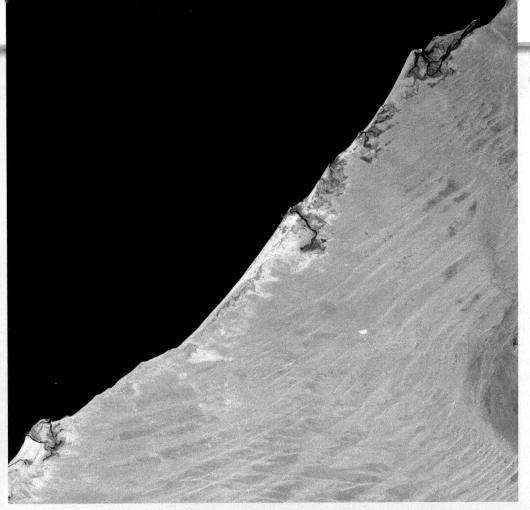

D Satellite image of Dubai, 24 January 1973, Landsat 1

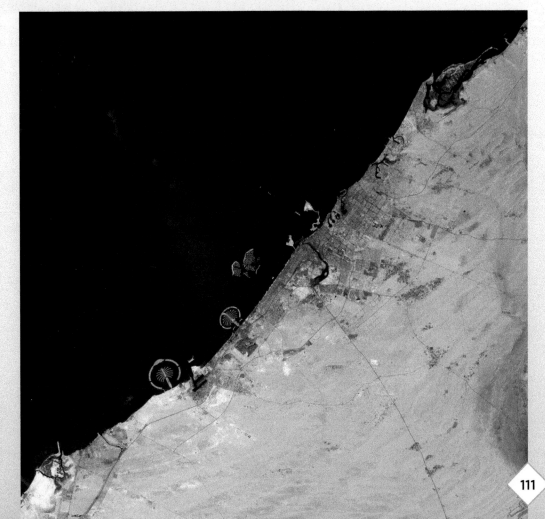

E Satellite image of Dubai, 11 October 2006, Landsat 7

Enquiry skills: How can I investigate how the UAE has changed? Part 2

So far in this enquiry you have investigated how the UAE and Dubai in particular has grown and transformed into a major city. The UAE enjoys a strategic location between Asia, Europe and Africa. It has used this advantage to transform itself into a world hub, developing airports, airlines, high-tech industries and tourism, to diversify the economy beyond oil and create wealth and jobs. The population of the UAE has grown rapidly since the 1970s, as people have been attracted to its modern lifestyle and job opportunities. Eighty-six per cent of the population live in the new cities like Dubai in the north-eastern corner of the country (map C on page 110), avoiding the vast deserts. Only 1.4 million people in the UAE are Emirati; 80 per cent of the population are migrants, attracted from all over the world to the job opportunities provided by the rapid economic growth. The UAE's expansion is totally reliant on this migrant workforce. In this lesson, you will focus your enquiry on the changes to the population of the country.

F A gridded population cartogram of the UAE, Worldmapper

G Population figures for the UAE, 1965–2019

Year	Population
1965	149,857
1970	234,514
1975	548,301
1980	1,019,509
1985	1,366,164
1990	1,828,432
1995	2,415,090
2000	3,134,062
2005	4,588,225
2010	8,549,988
2015	9,262,900
2019	9,770,529

H The population structure of the UAE, 2019

Age group, years	Percentage of total population	
	Males	Females
100+	0%	0%
95–99	0%	0%
90–94	0%	0%
85–89	0%	0%
80–84	0%	0%
75–79	0.1%	0%
70–74	0.4%	0.1%
65–69	0.7%	0.2%
60–64	1.7%	0.4%
55–59	3.5%	0.7%
50–54	5%	1.2%
45–49	7.7%	1.8%
40–44	8.7%	2.5%
35–39	8.9%	3.2%
30–34	10.4%	3.7%
25–29	9.2%	3.2%
20–24	5.6%	2.2%
15–19	2.7%	1.8%
10–14	2.2 %	2.1%
5–9	2.5%	2.4%
0–4	2.5%	2.4%

Activities

1 Look back at Lesson 1.1 to remind yourself of the enquiry process.
   a) The main question for this investigation of the UAE is in the title of this lesson – how has it changed? Identify two other enquiry questions that are important for you to consider as part of this enquiry.
   b) Identify and list the stages of your enquiry.

2 Look back at Lesson 1.4 and use map C (Lesson 14.1) and the guidance about describing the absolute and relative location of places. Write a paragraph to describe where the UAE is located.

3 Go to the http://confluence.org/ website.
   a) Search by countries and find the UAE. Go to each confluence point that has been visited.
   b) Write a paragraph to describe what the locations in the UAE are like and record the coordinates of each place.
   c) Download a photograph from one of the sites, print a copy and label its location and human and physical features.

4 Go to Google Earth.
   a) Search for 'UAE'. Zoom in and out of the country, and select 'Street View'. Use map C as a locator map.
   b) List five things you discover about the UAE.

5 Compare photos A and B and satellite images D and E.
   a) Compile a list of how Dubai has changed since the 1970s.
   b) Go to the ArcGiS storymap two-minute time lapse of Dubai. Add three more things to your list from 5a).

6 Go to the Dubai tourist website. Write a paragraph to explain how the city promotes itself as a world tourist destination.

7 Write a paragraph to explain how the UAE has developed.

8 Look carefully at map F and compare it with map C.
   a) What sort of map is map F?
   b) What is it showing about the UAE?
   c) How does the shape of the country in map F differ to the shape in map C?
   d) Write a paragraph to describe and explain the distribution patterns shown in map F.

9 Look at the data in table G.
   a) Decide which graph or chart would be best to represent the population change in the UAE from 1965 to 2019, and draw it.
   b) Write a paragraph to describe the population change.
   c) Identify three reasons for this population change and add them to your paragraph.

10 Look at the data in table H.
   a) Decide which graph or chart would be best to represent the population figures for the UAE, and draw it.
   b) Compare the shape of your completed chart for the UAE with those in Lesson 8.3. Identify ways that it is different.
   c) Think about how the UAE's population and economy have changed to identify why the pyramid for UAE is different.

11 Go to the population pyramid website for the UAE: www.populationpyramid.net/united-arab-emirates/2020/
   a) Look at the line graph for population growth for the UAE next to the pyramid. When you click on the line on the graph, the population pyramid changes to show the population structure for that year. Click on 1950 and then every ten years to 2019.
   b) Describe how the shape of the pyramid and age categories change.
   c) Continue to click on the line graph and observe how demographers predict that the UAE's population will change in the future.
   d) Describe what demographers think will happen to the population and the migrant workers of the UAE from 2060 onwards.

12 Write a conclusion to your enquiry to answer the title enquiry question for the lesson.

13 Look at the vision statement, flap C. Which aspects of this vision have you made progress in through conducting this enquiry about the UAE?

How can I draw a flow line map as part of an investigation of migrant workers in the UAE?

Learning objectives

▶ To be able to draw a flow line map to show the origin of migrant workers in the UAE.

▶ To use a newspaper article to consider issues for migrants.

A flow line map is a good way to show the volume of movement between places. Flow line maps can also be used to show the movement of exports and imports, traffic, and pedestrian or tourist movements. Flow line maps need to be drawn carefully to ensure they are not too cluttered, and do not obscure the detail of the map and different flows. The scale for the thickness of the arrows needs to be considered carefully to ensure all the flow lines for the map can be seen clearly.

Map A shows how to draw a flow line map. Table B shows the main countries of origin of migrant workers in the UAE. The number of migrants is shown by the varying thickness of the lines. The point of the arrow shows the direction of movement from the country of origin to the UAE. Only draw arrows for the major countries of origin, add the total numbers from other countries in the key to your map.

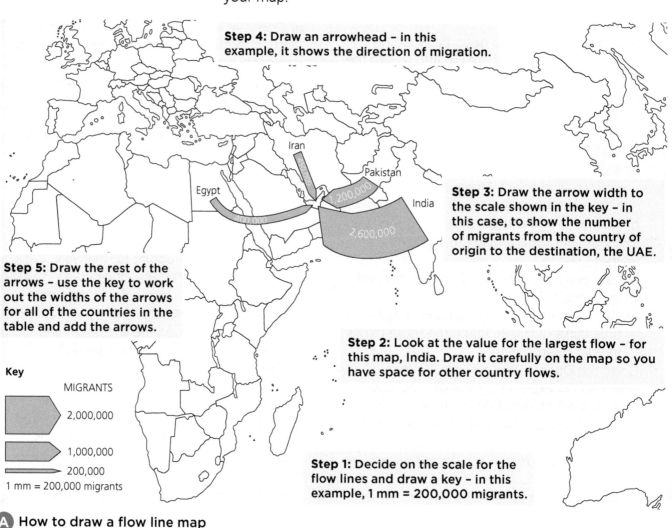

Step 4: Draw an arrowhead – in this example, it shows the direction of migration.

Step 3: Draw the arrow width to the scale shown in the key – in this case, to show the number of migrants from the country of origin to the destination, the UAE.

Step 5: Draw the rest of the arrows – use the key to work out the widths of the arrows for all of the countries in the table and add the arrows.

Step 2: Look at the value for the largest flow – for this map, India. Draw it carefully on the map so you have space for other country flows.

Step 1: Decide on the scale for the flow lines and draw a key – in this example, 1 mm = 200,000 migrants.

Iran
Pakistan
Egypt
India
1,200,000
400,000
2,600,000

Key

MIGRANTS

2,000,000

1,000,000

200,000

1 mm = 200,000 migrants

A How to draw a flow line map

Country of origin of migrants	Number of migrants
India	2,600,000
Pakistan	1,200,000
Bangladesh	700,000
Philippines	525,530
Iran	450,000
Egypt	400,000
Nepal	300,000
Sri Lanka	300,000
Syria	250,000
UK	250,000
China	200,000
Jordan	200,000
Afghanistan	150,000
Palestine	150,000
South Africa	100,000
Lebanon	100,000
Ethiopia	90,000
Yemen	90,000
Others	695,466

B Origin of migrants to the UAE
(the UAE National Bureau of Statistics does not publish demographic data in relation to any nationality. The figures listed in the table below are estimates provided by each country's embassy)

The UAE: a lure for low- and high-skilled migrants

The UAE attracts both low- and high-skilled migrants due to its economic attractiveness, relative political stability and modern infrastructure. Immigration has been the primary driver of population growth in the UAE since the 1990s, with immigrants making up the vast majority of the total population in 2019. Immigrants from Asia and the Middle East and North African region have dominated the low- and semi-skilled sectors, while workers from the United Kingdom, United States, Australia, Canada and various Western European countries have become concentrated in the UAE's key high-skilled sectors: the oil and gas industry, as well as banking and finance.

There are at least 400,000 foreign-born domestic workers in the UAE from mainly Asian countries. About 65,000 unauthorised migrants – including those who entered the country illegally, visa over-stayers, migrants working on tourist visas, and others – currently reside in the UAE, according to official estimates (unofficial estimates run up to 135,000).

C Extract from migration policy report, 2013

Activities

1 What is a flow line map?
2 What sort of data can be presented using flow line maps?
3 How does a flow line map show the volume of movement between two places?
4 Look carefully at map A and table B. The flows for four countries have been plotted on the map to help show you how to draw a flow line map.
   a) On an outline map of the world, create your own flow line map to show the data, using the guidance.
   b) Use map B in Lesson 1.4 to help you identify the countries of origin.
   c) Draw the flows using the scale provided on map A.
   d) Write a paragraph to describe the volume and distribution of the migrants' countries of origin.
5 Read the extract from a migration report, C.
   a) Add an evaluation to your paragraph in Question 4d) to explain why the data in table B might not be accurate.

   b) Compare your flow line map with map A from Lesson 7.1. Categorise the countries of origin of the migrants to the UAE into the GNI per capita groupings.
   c) Write a paragraph to explain why you think these people migrate to the UAE.
6 Conduct a search on the internet with 'news story treatment of migrant workers in UAE.' List 5 things your discover.
7 Reflect on what you have discovered about migrant workers in Lessons 14.1–14.3. Write a paragraph as a conclusion about migrant workers in the UAE. Include your answers to the following questions:
   a) Why are so many workers pulled to the UAE to work?
   b) Where is the migrants' place of origin, and what might be pushing them to migrate?
   c) What problems do they experience in the UAE?
   d) What might happen to migrant workers in the UAE in the future?

How can I use quantitative data to investigate climate change?

Learning objectives

▶ To interpret statistical data to investigate evidence of climate change.

▶ To compare data to reach a conclusion consistent with the evidence.

The climate has changed many times during Earth's history, but the changes have occurred slowly, over thousands of years. Scientific evidence indicates that the world's climate is now changing more quickly. As of 2018, the 20 warmest years on record globally have been in the past 22 years.

Ninety-seven per cent of the world's scientists now believe that since the mid-1800s, when humans began to burn fossil fuels such as coal, oil and gas for fuel, the release of carbon dioxide and other greenhouse gases into the atmosphere has increased (see B). These gases make up only one per cent of the atmosphere; they act like a blanket around the Earth, or like the glass roof of a greenhouse – they trap in the heat of the Sun that reflects back into space. This process is important as the trapped heat keeps the planet warm enough to sustain life. The increase in greenhouse gases, however, leads to more heat being trapped and an increase in global temperatures, affecting the delicate balance of the Earth's systems. In this lesson, you will investigate quantitative evidence of these changes.

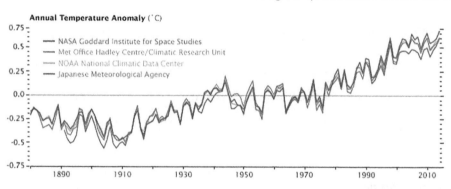

A Annual world temperature anomaly – the term *temperature anomaly* means a difference from the long-term average

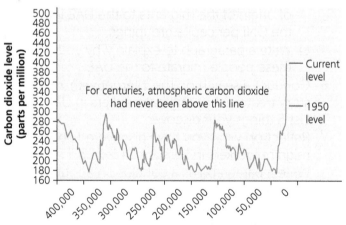

B Changing global levels of carbon dioxide

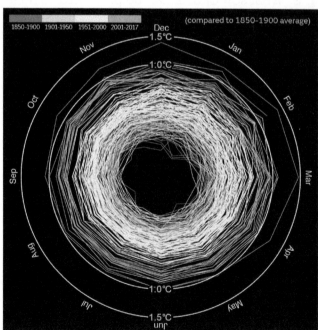

C Average global temperature anomaly by month, the Met Office

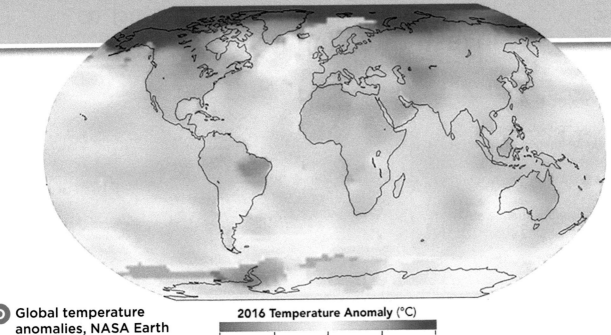

D Global temperature anomalies, NASA Earth Observatory

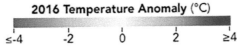

2016 Temperature Anomaly (°C)

≤-4   -2   0   2   ≥4

The global impact of climate change

Map D is taken from NASA's Earth Observatory website. As the map shows, global temperatures do not increase everywhere at the same rate.

The impact of global warming is far greater than just increasing temperatures. Warming changes rainfall patterns, increases coastal erosion due to a rise in sea level, lengthens the growing season in some regions, and melts ice caps and glaciers. Some of these changes are already occurring.

Activities

1 Look carefully at graph A.
   a) What is a temperature anomaly?
   b) Name the four different organisations that recorded yearly global temperature anomalies plotted on the graph.
   c) Write a paragraph to describe what the graph shows about world temperatures, recorded by these organisations.
   d) How does the fact that each organisation recorded similar results make the evidence more reliable?

2 Look carefully at graph C.
   a) Write a paragraph to describe what the graph shows.
   b) How much has the average global temperature increased since 1850?
   c) Identify three things that this graph adds to your understanding of how global temperatures are changing.

3 Look at map D.
   a) Write three sentences about what the map shows about global temperatures in 2016.
   b) Identify which parts of the world had the largest temperature anomaly.
   c) Do you think that the effects of increased global temperatures will only affect the part of the world with the highest temperature anomaly? Explain your answer.

4 Look at graph B.
   a) Describe how the levels of carbon dioxide in the atmosphere have changed.
   b) Go to the NASA website: https://svs.gsfc. nasa.gov/11719. Watch the animation, and explain what happens to carbon dioxide in the atmosphere.

5 Write a conclusion to summarise what you have learnt about climate change.

Learning objectives

▶ To understand how geographers can use satellite imagery to monitor change.

▶ To interpret satellite images to identify glacial retreat.

▶ To conduct an independent web enquiry about changing glaciers.

In order to prove the existence of climate change, scientists are continually researching ways to measure the rate at which the planet's temperature is increasing. One way they can do this is to use satellites to monitor heat-sensitive objects on the ground, such as glaciers. Small glaciers tend to respond more quickly to climate change than the giant ice sheets. However, as no two glaciers are alike, scientists have to study a wide range of glaciers across different regions to draw conclusions.

NASA's Earth Observatory website has archived data of glaciers and ice sheets which we can use to investigate climate change. The Columbia Glacier is one such example. Descending from an ice field 3050 metres above sea level, down the sides of the Chugach Mountains, the Columbia Glacier ends in a narrow inlet leading into Prince William Sound (see map A). It is one of the most rapidly changing glaciers in the world. When British explorers first surveyed it in 1794, its snout extended to the northern edge of Heather Island. The glacier held that position until 1980, when it began a rapid retreat, which continues today. The false-colour images, B–D, were captured by Landsat. The 2011 image has more snow because it was captured in May, while the 1986 image was captured in July and the 2019 image in June. By 2011, the Columbia Glacier had retreated more than 20 kilometres and had split the Columbia into two glaciers.

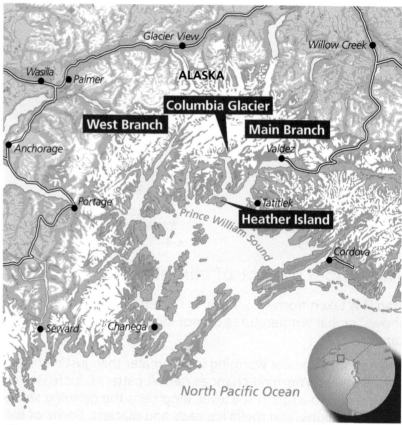

A Map of the Columbia Glacier in South East Alaska

Activities

1 Why are scientists trying to predict the rate of global temperature increase using satellites?

2 Why are glaciers a good indicator of climate change?

3 Go to Google Earth and find the area shown in map A.
   a) Locate the Columbia Glacier.
   b) Describe the location of the Columbia Glacier.

4 Compare map A with satellite image B. Name locations 1–3.

5 Compare satellite images B–D.
   a) Write a paragraph describing how the glacier has changed from 1986 to 2019.
   b) Write three sentences to explain what you think this suggests about climate change.

6 Use the NASA Earth Observatory website – https://earthobservatory.nasa.gov – to investigate how other glaciers and ice sheets are changing.

7 What impact do you think melting glaciers and ice sheets will have on sea levels?

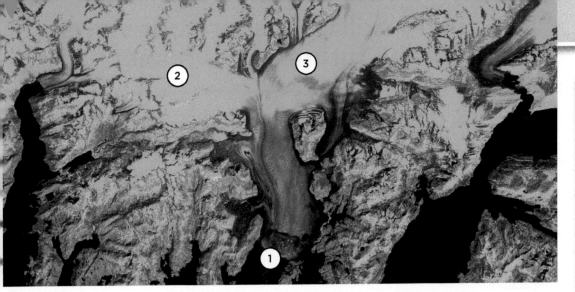

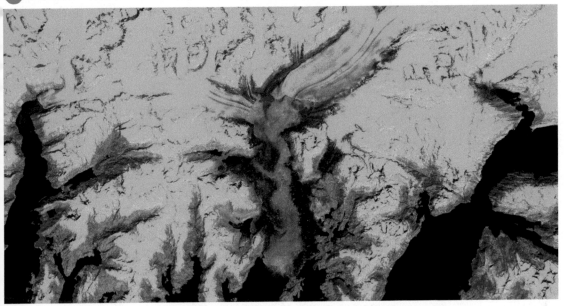

B Columbia Glacier, 28 July 1986, Landsat 5, NASA Earth Observatory

C Columbia Glacier, 30 May 2011, Landsat 5, NASA Earth Observatory

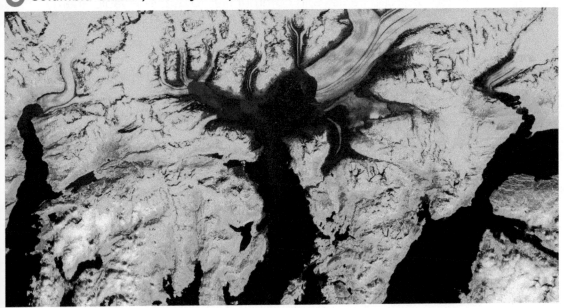

D Columbia Glacier, 21 June 2019, Landsat 8, NASA Earth Observatory

Enquiry skills: What are our views about the future of the planet?

Learning objectives

▶ To consider different viewpoints about climate change.

▶ To classify viewpoints.

▶ To discuss a global issue and justify your viewpoint.

▶ To consider the views of others to reach your own decision about future actions.

The consequences of climate change zoom through the scales of the Earth from global to local and individual. It is often difficult to relate events of extreme weather to slowly emerging global changes in the world's temperatures. The scientific evidence and global view of the planet provided by satellites present a compelling overview of the change. This, together with the increasing frequency of extreme weather events and rising sea levels as glaciers and ice sheets start to melt, have all made it clearer that climate change is an issue affecting the future of the planet.

In this lesson, you will consider different opinions about climate change, to determine which option is being represented and at what scale, before discussing your own views, to make a class decision about the future of the planet.

Possible action

Many now believe the future of the planet will be determined by three possible options from this point forward:

1 **Do nothing** and live with the consequences.

2 **Adapt** to the changing climate, adjusting to the rising sea level and extreme weather.

3 **Mitigate** – make the consequences of climate change less severe, by finding ways to reduce the concentration of greenhouse gases in the atmosphere.

Actions at different scales

Many now believe we need to use the planet in a more sustainable way, meeting the needs of the present without compromising the well-being of future generations.
This is an important concept for you to understand as a geographer, and as a responsible citizen. Actions to improve sustainability can operate at different scales:

● **Local** – by individuals, schools, and communities, for example recycling material as part of waste disposal or saving energy and reducing carbon emissions by using cars less.

● **National** – the UK government has begun to encourage sustainable use of energy by offering incentives to companies and communities to use renewable energy.

● **International** – organisations attempting to get a wide range of countries to work together to tackle the global issue of climate change.

Activities

1 Read the views expressed in A–H.

   a) Work with a partner and categorise these views into the three options facing the planet, and the three different scales (see textboxes above).

   b) Identify three views you agree with and three you do not. In each case, explain your choice.

2 Your class will be divided into three groups. Each group will prepare a case for one of the options for action facing the planet, using the viewpoints as a starting point.

   a) Prepare a presentation to represent the views of your group for your option.

   b) Present your group's views to the rest of the class.

   c) Debate the three options.

   d) The class can conclude by making a decision about the best option for the future.

   e) Write a paragraph to justify the decision made using the evidence from the debate and views A–H.

The long-term goal approved by 195 countries is to keep the increase in global average temperature to well below 2°C above pre-industrial levels; to limit the increase to 1.5°C.

A Paris Climate Accord

The climate crisis is both the easiest and the hardest issue we have ever faced. The easiest because we know what we must do. We must stop the emissions of greenhouse gases. The hardest because our current economics are still totally dependent on burning fossil fuels, and thereby destroying ecosystems in order to create everlasting economic growth. Humans are very adaptable: we can still fix this. But the opportunity to do so will not last for long. We must start today. We have no more excuses.

C Sixteen-year-old Swedish climate activist Greta Thunberg's speech to MPs at the Houses of Parliament, 23 April 2019

The UK led the world to wealth through fossil fuels in the industrial revolution, so it was appropriate for Britain to lead in the opposite direction. We have reduced our coal dependency from 40 per cent in 2012 to less than 2 per cent. As a result, in 2017, we had the first full day that the UK has not used coal for 135 years. The UK already has a 2050 target – to reduce emissions by 80 per cent. That was agreed by MPs under the Climate Change Act in 2008, but will now be amended to the new, much tougher, goal committing the UK to net zero emissions by 2050.

B Theresa May, former UK Prime Minister, June 2019

I try to set up my electric appliances differently, unplug them whenever I can, so they are not connected to the grid unnecessarily, using power. Government seems to blame individuals or us rather than the big companies that create massive carbon emissions. We need to target them to do better. To mitigate against climate change we need to get off using fossil fuels and transition to renewables. I think a lot of older people think I'll be dead before climate change affects me, but the youth are the people who really have to worry about it.

The warming had already started by the 1930s. That's when there were no such factors as carbon emissions, and the warming had already started. The issue is not stopping it … because that's impossible since it could be tied to some global cycles on Earth or even of planetary significance. The issue is to somehow adapt to it. The melting Arctic could bring more favourable conditions for using this region for economic ends, exposing natural resources and transport routes which had long been too expensive to exploit.

F World youth climate change protest, 7 March 2015

H The first of Iceland's 400 glaciers to be lost to the climate crisis remembered with a memorial plaque – and a sombre warning for the future

D Russian President Vladimir Putin, 31 March 2017, visiting the Franz Josef Land archipelago in the Arctic

As sea levels rise and storms get stronger, we are nationally struggling with what to do with our crumbling sea defences. The assumption that bigger, stronger, higher seawalls and defences can hold back the waves has created an endless cycle of expensive repair and construction. Protecting historic and new development on the shoreline pits the irresistible force of the sea against the immovable object of steel and concrete and it's a fight neither side can ultimately win.

E Stewart Rowe, Principal Coastal Officer, Scarborough Borough Council

In order to fulfil my solemn duty to protect America and its citizens, the United States will withdraw from the Paris Climate Accord. The cost to the economy of the Accord at this time would be close to $3 trillion in lost GDP and 6.5 million industrial jobs, while households would have $7000 less income. We have among the most abundant energy reserves on the planet, sufficient to lift millions of America's poorest workers out of poverty. Yet, under this agreement, we are effectively putting these reserves under lock and key, taking away the great wealth of our nation.

G US President Donald Trump on the Paris Climate Accord

Progress in Geography Skills review

In **Progress in Geography Skills**, you have learnt:

▶ about being a geographer
▶ how to develop and use a wide range of geographical skills
▶ how to conduct geographical enquiries
▶ how to ask geographical questions
▶ how to use a wide range of geographical data, including OS maps.

These were the key objectives introduced and progressed in **Progress in Geography Skills**. You have now completed this stage of the journey of becoming a geographer.

Let's see what you've learnt and how this connects to the study of Geography at GCSE level.

Activities

1 In Lesson 1.1, you were introduced to the geographical enquiry process. Think back about your Key Stage 3 Geography studies, and write a paragraph to describe how you conduct such enquiries.

2 In Lesson 1.1, you were also asked to create a table of the types of geographical data provided in this textbook. Now you have completed the course, draw a table like the one in Lesson 1.1, Question 1, and this time list examples of how you have used each type of geographical data during your studies in Key Stage 3. (Leave space on your table to add columns for Question 4.)

3 Look carefully at the vision statement for **Progress in Geography Skills** on flap C. Identify and write down which opportunities developed your geographical skills. For each, give an example of how you made progress.

4 Read table B, which shows a list of skills for the AQA GCSE Geography course.

a) Add two columns to your table from Question 2 – one with the heading 'Skills to progress further at GCSE' and the other with 'New geography skills at GCSE'.

b) Now fill in these two columns.

c) Use your table to write a paragraph to summarise how you think your progress in developing geographical skills has prepared you for GCSE Geography.

5 Each year after the GCSE examinations the Chief Examiner writes a report for schools summarising how students performed in the exam, and lessons that can be learnt from this. Quotes A and C are taken from some of these reports to demonstrate the importance of geographical skills at GCSE.

Read these quotes carefully, and then write a paragraph to explain why it is important that you continue to progress your geographical skills in your future studies.

> Where a source is provided, students are expected to make use of it as a springboard to develop an answer, particularly where the command is 'using figure … and your own knowledge/understanding'. … how to respond to statistical data, different types of graph and a range of maps at different scales is important … allowing students to interpret and analyse the information, deal with patterns, and comment on trends and anomalies. A sizeable number struggled to make full use of the geographical sources provided and gave vague generic responses or misunderstood the main focus of the question. Photographic evidence in particular was not always carefully considered and linked into the student's answer.

Ⓐ **AQA GCSE Geography Examiner's Report, Paper 1 2018**

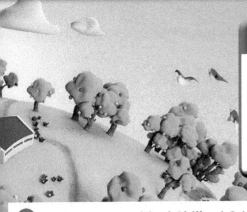

Future learning at GCSE

As you have discovered, *Progress in Geography Skills: Key Stage 3* has been designed to help you become a good geographer and provide you with a foundation of skills for you to build on to make progress and succeed at GCSE. Table B summarises the skills students are required to develop at GCSE level. These skills will be assessed in all three written exams.

B GCSE Geographical Skills, AQA specification (2016), pages 29–31

Geography skill category	Geographical skills
Cartographic skills – atlas maps	Use and understand coordinates – latitude and longitude; recognise and describe distribution patterns of human and physical features, e.g. population distribution, population movements, transport networks, settlement layout, relief and drainage; analyse the inter-relationship between physical and human factors on maps and establish associations between observed patterns on thematic maps.
Cartographic skills – OS maps	Use and interpret OS maps at a range of scales, including 1:50 000 and 1:25 000 and other maps appropriate to the topic; four- and six-figure grid references; use and understand scale, distance and direction, measure straight and curved line distances using a variety of scales; gradient, contour and spot height; identify basic landscape features and describe their characteristics from map evidence, interpret cross sections and transects.
Maps in association with photographs	Be able to compare maps; sketch maps: draw, label, understand and interpret; photographs: use and interpret ground, aerial and satellite photographs; describe human and physical landscapes (landforms, natural vegetation, land-use and settlement) and geographical phenomena from photographs; draw sketches from photographs; label and annotate diagrams, maps, graphs, sketches and photographs.
Graphical skills	Select and construct appropriate graphs and charts to present data, using appropriate scales – line charts, bar charts, pie charts, pictograms, histograms with equal class intervals, divided bar, scatter graphs, and population pyramids; suggest an appropriate form of graphical representation for the data provided; complete a variety of graphs and maps – choropleth, isoline, dot maps, desire lines, proportional symbols and flow lines; use and understand gradient, contour and value on isoline maps; plot information on graphs when axes and scales are provided; interpret and extract information from different types of maps, graphs and charts, including population pyramids, choropleth maps, flow-line maps and dispersion graphs.
Numerical skills	Demonstrate an understanding of number, area and scales, and the quantitative relationships between units; design fieldwork data collection sheets and collect data with an understanding of accuracy, sample size and procedures, control groups and reliability; understand and correctly use proportion and ratio, magnitude and frequency; draw informed conclusions from numerical data.
Statistical skills	Use appropriate measures of central tendency, spread and cumulative frequency (median, mean, range, quartiles and inter-quartile range, mode and modal class); calculate percentage increase or decrease and understand the use of percentiles; describe relationships in bivariate data: sketch trend lines through scatter plots, draw estimated lines of best fit, make predictions, interpolate and extrapolate trends; be able to identify weaknesses in selective statistical presentation of data.
Use of qualitative and quantitative data	Use qualitative and quantitative data from both primary and secondary sources to obtain, illustrate, communicate, interpret, analyse and evaluate geographical information. Examples of types of data: maps, fieldwork data, geo-spatial data presented in a GIS, satellite imagery, written and digital sources, visual and graphical sources.
Formulation of enquiry and argument	Identify questions and sequences of enquiry; write descriptively, analytically and critically; communicate their ideas effectively; develop an extended written argument; draw well-evidenced and informed conclusions about geographical questions and issues.

When describing resources such as maps and graphs, candidates should make use of the information provided. Accurate reference to data, scale and compass directions will gain credit in the context of the question.

C Eduqas A GCSE Geography Examiner's Report, 2018

How will I use geographical skills in the future?

Learning objectives

▶ To appreciate how people use geographical skills in their lives.

▶ To consider how I may use geographical skills in my future.

The national curriculum for Geography states that 'a high-quality geography education should inspire in pupils a curiosity and fascination about the world and its people that will remain with them for the rest of their lives'. The geographical skills you have developed in Key Stage 3 are skills that you will utilise and further develop throughout your life, as you interact with people, places and environments around the world, both in your work and leisure time. The following people are former Geography pupils of the author of this textbook, who reflect on how they are using these skills in their adult life.

As a Navigational Deck Officer on board large cruise ships, I use geographical skills every day. I plan routes to various ports across the world, using maps and charts to navigate across seas and oceans, plotting the vessel position on charts using GPS and visual fixes from geographical landmarks, ensuring the ship is kept safe on passage. I also analyse synoptic charts to predict weather, tidal conditions, ocean currents and port information. The ship also acts as a weather station for the Met Office. I transmit weather data daily to them to support their analysis of daily global weather patterns used to predict the weather.

A James Buck, age 30, 3rd Officer (Ship Navigation)

While studying Geography in school at Key Stage 3 and GCSE, we conducted geographical enquiries each year. This taught me to ask interesting questions about the world, and how to decide which tools, data and presentation techniques to best answer these questions. I have developed and progressed these skills, which now inform my research work as an academic, working in a university, answering questions about environmental activism in China, carbon off-setting schemes and the lives of street children in Tanzania.

B Dr Gemma Pearson, age 33, Doctoral/ Academic Researcher in Geography

Travelling the world is a passion of mine, and I love to explore new places. In this photo, I'm in Istanbul near the River Bosphorus. Before I visit a new destination, I consult an atlas, photos, websites and travel guides to help me plan my route and daily itinerary. I also review climate data, graphs and weather charts and forecasts to ensure I am travelling in the most appropriate season.

C Helena Jaconelli, age 33, Senior Merchandiser

At school as part of geographical investigations we drew fieldsketches, which I really enjoyed. One of the bestselling lines in my shop are illustrations of Scarborough landscapes, which I now draw and print out as greetings cards, tea towels and framed pictures. In the future, I hope to draw a map of Scarborough incorporating my illustrations of local landmarks. On a recent buying trip to India, I used a variety of maps, both paper and digital, to source manufacturers to develop my furniture designs.

D **Alex Anderson, age 35, Company Owner/Director, interior design shop**

I use the skills I learnt in KS3 Geography on a daily basis. My job as a Chartered Surveyor involves the disposal and purchase of properties, in particular warehouse/distribution units. Their location is one of the most important factors in determining value and suitability of properties. Factors such as proximity to motorway junctions, for ease of access, or major conurbations, in order to draw on labour resources, are considered. We use mapping software at a variety of scales, to analyse buildings based on their location, but also to establish information such as area and building sizes.

E **Thomas Goode, age 32, Chartered Surveyor**

I've always had a keen interest in Geography and went on to read it at university. The wide variety of skills and breadth of study in Geography, from human to physical and all in-between, has helped me throughout my studies and into the world of work. Studying these elements has helped me significantly in my present role, focusing on the public and how they interact with the environment – from the waste they produce to the parks they visit, how areas can be regenerated and the impact the environment has on people's health.

F **Harry Briggs, age 35, Deputy Operations, Transport and Countryside Manager, Scarborough Borough Council**

Activities

1. Read the views of the former Geography pupils (A–F).
   a) In each case, describe how they are using geographical skills in their everyday adult life or work.
   b) In each case, explain how they are using and progressing the geographical skills you, and they, have studied in Key Stage 3.
   c) Compare what you have discovered about the lives of each person with the vision statement, flap C. Write a paragraph to explain why you think they are excellent geographers.

2. Look back at the geographical skills introduced in this book. Write a paragraph to describe which skills you think you might develop and use in the future.

Index

Acknowledgements

Photo credits

p.2, p.4 © David Gardner; **p.5** © Kevin KAL Kallaugher, The Economist, Kaltoons.com; **pp.12–15** © David Gardner; **p.19** © Getmapping PLC; **pp.27–29** © David Gardner; **p.34** Sirius Minerals Plc; **p.41** © David Gardner; **p.43, p.45** Reproduced with the permission of University of Dundee Archives; **p.56** © geogphotos/Alamy Stock Photo; **p.57** © Getmapping PLC; **p.58** © EUMETSAT 2020, Meteosat-10 Airmass RGB, 05 Dec 00:00 UTC; **p.59** © Crown copyright, Met Office; **p.60** Environment Agency; **p.66** © Hodder & Stoughton; **p.67** © Jann Lipka/gapminder.org/https://creativecommons.org/licenses/by/4.0; **p.68** © Vinap/stock.adobe.com; **p.75** © Commission Air/Alamy Stock Photo; **p.76** © David Gardner; **p.77** © RiRi/Stockimo/Alamy Stock Photo; **pp.78–79** © The National Trust Photolibrary/Alamy Stock Photo; **p.80** © David Gardner; **p.81** *l* Courtesy of Stewart Rowe; *r* © David Gardner; **p.83** © Paul White-Real Yorkshire/Alamy Stock Photo; **p.87** NASA Earth Observatory images by Lauren Dauphin, using Landsat data from the U.S. Geological Survey; **p.90** Antonio Violi/Alamy Stock Photo; **p.91** *l* © Oleksandr Prykhodko/Alamy Stock Photo; *r* © David Gardner; **p.92** Reproduced with the permission of University of Dundee Archives; **p.93** *t* NASA Earth Observatory images by Joshua Stevens, using Landsat data from the U.S. Geological Survey; **p.98** NASA Earth Observatory image using Landsat data; **p.102** © David Gardner; **p.105** © Getmapping PLC; **p.107** © David Gardner; **p.110** *l* © Art Directors & TRIP/Alamy Stock Photo; *r* © Luciano Mortula/Alamy Stock Photo; **p.111** NASA image created by Jesse Allen, Earth Observatory, using data provided by Laura Rocchio, Landsat Project Science Office; **p.119** *t and c* NASA images by Jesse Allen and Robert Simmon, using Landsat 5 data from the USGS Global Visualization Viewer; *b* NASA Earth Observatory images by Lauren Dauphin, using Landsat data from the U.S. Geological Survey; **p.121** *tr* © Shutterstock/Michael Tubi; *tl* © dpa picture alliance/Alamy Stock Photo; *c* © Bjanka Kadic/Alamy Stock Photo; *cr* © JEREMIE RICHARD/AFP via Getty Images; *br* © MediaPunch Inc/Alamy Stock Photo; *bl* © Planetpix/Alamy Stock Photo; *b* Courtesy of Stewart Rowe; **pp.122–123** © arquiplay77 - stock.adobe.com; **p.124** *t* Courtesy of James Buck; *c* Courtesy of Dr Gemma Pearson; *b* Courtesy of Helena Jaconelli; **p.125** *t* Courtesy of Alex Anderson; *c* Courtesy of Thomas Goode; *b* Courtesy of Harry Briggs.

Text credits

Maps included by kind permission of Ordnance Survey (OS) on **pp.10, 11, 26, 28, 32–33, 57, 102** and cover.

Atlas maps included by kind permission of Philip's on **pp.4, 6–7, 8–9, 30, 31, 48–49, 84, 85, 96** and **97**.

p.23 © Liverpool City Council 2015, Map updated October 2019; **p.24** © Southwark Council; **p.32–33** © Sirius Minerals Plc; **p.39** © Port of Liverpool map, Peel Ports Group; **p.40** © Dorset Council; **p.50** confluence.org © 2008 by Alex Jarrett; **p.60** Likelihood of flooding in this area © Crown copyright. Contains public sector information licensed under the Open Government Licence v3.0; **p.63** GDP Wealth 2018 © Worldmapper; **p.64** Gapminder Bubble chart tool guide © Gapminder; **p.66** Dollar Street © Gapminder; **p.69** Population Year 2018 © Worldmapper; **p.88** What is China's Belt and Road Initiative? © Guardian News & Media Limited; **p.89** Yu Hongjun, China's integration with world will stay on course © Chinadaily.com; **p.93** In-depth study and update on seismic and volcanic activity in the Etna area © blogingvterremoti; **p.94** USGS GIS earthquake map tool guide © USGS; **p.99** UNESCO launches BIOPALT project to safeguard Lake Chad, Monday, 5 March 2018 © UNESCO; **p.109** BRITICE Glacial Map v2.0 © The University of Sheffield; **p.112** United Arab Emirates Gridded Population © Worldmapper; **p.116** *tl* World of Change: Global Temperatures © NASA; *br* What is climate change? MetOffice, © Crown Copyright. Contains public sector information licensed under the Open Government Licence v3.0; **p.117** Global Temperature Record Broken for Third Consecutive Year © NASA.

Legend for Ordnance Survey 1:50 000 maps

Communications

ROADS AND PATHS

Not necessarily rights of way

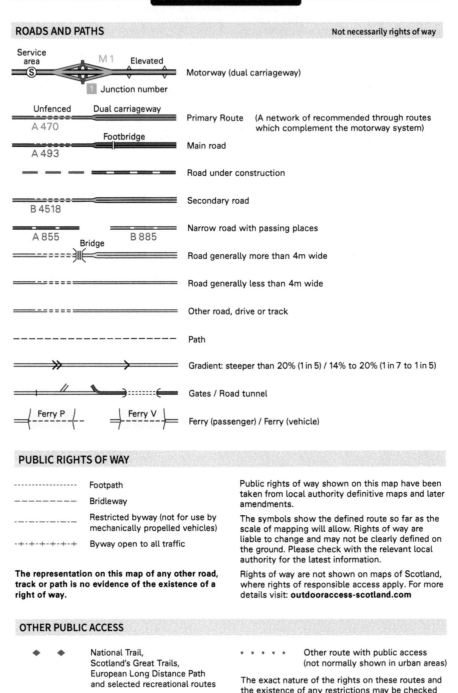

Motorway (dual carriageway)

Service area (S)
M 1
Elevated
1 Junction number

Unfenced — Dual carriageway — Primary Route (A network of recommended through routes which complement the motorway system)
A 470

Footbridge
A 493 — Main road

Road under construction

B 4518 — Secondary road

A 855 — B 885 — Narrow road with passing places

Bridge
Road generally more than 4m wide

Road generally less than 4m wide

Other road, drive or track

Path

Gradient: steeper than 20% (1 in 5) / 14% to 20% (1 in 7 to 1 in 5)

Gates / Road tunnel

Ferry P — Ferry V — Ferry (passenger) / Ferry (vehicle)

PUBLIC RIGHTS OF WAY

---------------- Footpath

– – – – – – – Bridleway

–·–·–·–·– Restricted byway (not for use by mechanically propelled vehicles)

-+-+-+-+-+ Byway open to all traffic

The representation on this map of any other road, track or path is no evidence of the existence of a right of way.

Public rights of way shown on this map have been taken from local authority definitive maps and later amendments.

The symbols show the defined route so far as the scale of mapping will allow. Rights of way are liable to change and may not be clearly defined on the ground. Please check with the relevant local authority for the latest information.

Rights of way are not shown on maps of Scotland, where rights of responsible access apply. For more details visit: **outdooraccess-scotland.com**

OTHER PUBLIC ACCESS

◆ ◆ National Trail, Scotland's Great Trails, European Long Distance Path and selected recreational routes

National Cycle Network;

● ● ● On-road cycle route

○ ○ ○ Traffic-free cycle route

4 8 Cycle route number; National / Regional

· · · · · Other route with public access (not normally shown in urban areas)

The exact nature of the rights on these routes and the existence of any restrictions may be checked with the local highway authority. Alignments are based on the best information available. These routes are not shown on maps of Scotland.

Danger Area
Firing and Test Ranges in the area. Danger! Observe warning notices.

RAILWAYS

Track multiple or single

Track under construction

Light rail system, narrow gauge or tramway

Bridges, footbridge

Tunnel, cutting

Station, (a) principal

Siding

Light rail system station

Level crossing LC

Viaduct, embankment

General Information

BOUNDARIES

National

District

County, Unitary Authority, Metropolitan District or London Borough

National Park

WATER FEATURES

Contour values in lakes are in metres

Marsh or salting

Towpath

Aqueduct

Canal

Weir

Footbridge   Bridge

Lake

Canal (dry)

Lock

Ford

Normal tidal limit

Slopes

Cliff

Flat rock

Sand Dunes

Mud

High water mark

Low water mark

Lighthouse (in use)

Lighthouse (disused)

Beacon

Shingle

LAND FEATURES

Cutting, embankment

Electricity transmission line (pylons shown at standard spacing)

Pipe line (arrow indicates direction of flow)

Buildings

Important building (selected)

Bus or coach station

Glass Structure

Heliport

Current or former place of worship; with tower
with spire, minaret or dome

Place of worship

Triangulation pillar

Mast

Wind pump

Wind turbine

Windmill with or without sails

Graticule intersection at 5' intervals

Landfill site or slag/spoil heap

Coniferous wood

Non-coniferous wood

Mixed wood

Orchard

Park or ornamental ground

Access land (symbols indicate owner or agency – see below)

Forestry Commission

Natural Resources Wales

National Trust; always open, limited access – observe local signs

National Trust for Scotland; always open, limited access – observe local signs

Forestry Division Plantation (Isle of Man)

Manx National Heritage

ABBREVIATIONS

See our website for full list

Br	Bridge	MS	Milestone
Cemy	Cemetery	Mus	Museum
CG	Cattle grid	P	Post office
CH	Clubhouse	PC	Public convenience (in rural areas)
Coll	College	PH	Public house
Fm	Farm	Sta	Station
Ho	House	Sch	School
Hospl	Hospital	TH	Town Hall, Guildhall or equivalent
MP	Milepost	Univ	University

CONVERSION

METRES – FEET

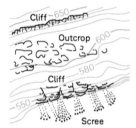

1 metre = 3.2808 feet

15.24 metres = 50 feet

ARCHAEOLOGICAL AND HISTORICAL INFORMATION

+ Site of antiquity   VILLA Roman   ⚔ Site of battle (with date)

☆ ᵐ Visible earthwork   𝕮𝖆𝖘𝖙𝖑𝖊 Non-Roman

Information sourced from Historic England, Historic Environment Scotland and the Royal Commission on the Ancient and Historical Monuments of Wales.

HEIGHTS

—50— Contours are at 10 metres vertical interval

·144 Heights are to the nearest metre above mean sea level

Where two heights are shown, the first is the height of the natural ground in the location of the triangulation pillar, and the second (in brackets) to a separate point which is the natural summit.

ROCK FEATURES

Cliff — 650
Outcrop — 600
Cliff — 580
— 550
Scree

Tourist Information

	Viewpoint 180°		Camp site / Caravan site
	Viewpoint 360°		Camping and caravan site
	Visitor centre		Selected places of tourist interest
	Walks / trails		Information centre, all year / seasonal
	Nature reserve		Parking
	Picnic site		Park & Ride, all year / seasonal
	Youth hostel		Phone, public / emergency
	Golf course or links		Recreation / leisure / sports centre
	Garden / arboretum		World Heritage site or area